Reports of China's Basic Research

Editor-in-Chief

Wei Yang, National Natural Science Foundation of China, Beijing, China, Zhejiang University, Hangzhou, Zhejiang, China

The National Natural Science Foundation of China (NSFC) was established on February 14, 1986. Upon its establishment, NSFC was an institution directly under the jurisdiction of the State Council, tasked with the administration of the National Natural Science Fund from the Central Government. In 2018, it became managed by the Ministry of Science and Technology (MOST) but kept its due independence in operation. Since its establishment, NSFC has comprehensively introduced and implemented a rigorous and objective merit-review system to fulfill its mission of supporting basic research, fostering talented researchers, developing international cooperation and promoting socioeconomic development.

Featuring science, basics, and advances, the series of *Reports of China's Basic Research* is organized by the NSFC to present the overall level and pattern of China's basic research, share innovative achievements, and illustrate excellent breakthroughs in key fields. It covers various disciplines including but not limited to, computer science, materials science, life sciences, engineering, environmental sciences, mathematics, and physics. The series will show the core contents of the final reports of the Major Programs and the Major Research Plans funded by NSFC, and will closely follow the frontiers of basic research developments in China.

If you are interested in publishing your book in the series, please contact Qian Xu (Email: violetta_xuqian@zju.edu.cn) and Mengchu Huang (Email: mengchu.huang@cn.springer.com).

Lihe Zhang
Editor

Investigations on Signal Transduction Processes Utilizing Small Chemical Probes

ZHEJIANG UNIVERSITY PRESS
浙江大学出版社

Springer

Editor
Lihe Zhang
Health Science Centrre
Peking University
Beijing, China

ISSN 2731-8907 ISSN 2731-8915 (electronic)
Reports of China's Basic Research
ISBN 978-981-19-9584-2 ISBN 978-981-19-9585-9 (eBook)
https://doi.org/10.1007/978-981-19-9585-9

Jointly published with Zhejiang University Press
The print edition is not for sale in China (Mainland). Customers from China (Mainland) please order the print book from: Zhejiang University Press.

Translation from the Chinese language edition: "Jiyu Huaxue Xiaofenzi Tanzhende Xinhao Zhuandao Guocheng Yanjiu" by Lihe Zhang, © Zhejiang University Press 2019. Published by Zhejiang University Press. All Rights Reserved.

This Springer imprint is published by the registered company Springer Nature Singapore Pte Ltd.
The registered company address is: 152 Beach Road, #21-01/04 Gateway East, Singapore 189721, Singapore

Reports of China's Basic Research

Preface to the Series

As Lao Tzu said, "A huge tree grows from a tiny seedling; a nine-storied tower rises from a heap of earth." Basic research is the fundamental approach to fostering innovation-driven development, and its level becomes an important yardstick for measuring the overall scientific and national strength of a country. Since the beginning of the twenty-first century, China's overall strength in basic research has been consistently increasing. With respect to input and output, China's input in basic research increased by 14.8 times from 5.22 billion yuan in 2001 to 82.29 billion yuan in 2016, with an average annual increase of 20.2%. In the same period, the number of China's scientific papers included in the Science Citation Index (SCI) increased from lower than 40,000 to 324,000; China rose from the 6th to 2nd place in global ranking in terms of the number of published papers. In regard to the quality of output, in 2016, China ranked No. 2 in the world in terms of citations in nine disciplines, among which materials science ranked No. 1; as of October 2017, China ranked No. 3 in the world in the numbers of both Highly Cited Papers (top 1%) and Hot Papers (top 0.1%), with the latter accounting for 25.1% of the global total. In talent cultivation, in 2006, China had 175 scientists (136 of whom from the Chinese mainland) included in Thomson Reuters' list of Highly Cited Researchers, ranking 4th globally and 1st in Asia.

Meanwhile, we should also be keenly aware that China's basic research is still facing great challenges. First, funding for basic research in China is still far less than that in developed countries—only about 5% of the R&D funds in China are used for basic research, a much lower percentage than 15%–20% in developed countries. Second, competence for original innovation in China is insufficient. Major original scientific achievements that have a global impact are still rare. Most of the scientific research projects are just a follow-up or imitation of existing research, rather than groundbreaking research. Third, the development of disciplines is not balanced, and China's research level in some disciplines is noticeably lower than the international level—China's Field-Weighted Citation Impact (FWCI) in disciplines just reached 0.94 in 2016, lower than the world average of 1.0.

The Chinese government attaches great importance to basic research. In the 13th Five-Year Plan (2016–2020), China has established scientific and technological innovation as a priority in all-round innovation and has made strategic arrangements to strengthen basic research. General Secretary XI Jinping put forward a grand blueprint for making China a world-leading power in science and technology in his speech delivered at the National Conference on Scientific and Technological Innovation in 2016 and emphasized that "we should aim for the frontiers of science and technology, strengthen basic research, and make major breakthroughs in pioneering basic research and groundbreaking and original innovations" at the 19th CPC National Congress on October 18, 2017. With more than 30 years of unremitting exploration, the National Natural Science Foundation of China (NSFC), one of the main channels for supporting basic research in China, has gradually shaped a funding pattern covering research, talent, tools, and convergence, and has taken action to vigorously promote basic frontier research and the growth of scientific research talent, reinforce the building of innovative research teams, deepen regional cooperation and exchanges, and push forward multidisciplinary convergence. As of 2016, nearly 70% of China's published scientific papers were funded by the NSFC, accounting for 1/9 of the total number of published papers all over the world. Facing the new strategic target of building China into a strong country in science and technology, the NSFC will conscientiously reinforce forward-looking planning and enhance the efficiency of evaluation, so as to achieve the strategic goal of making China progressively share the same level with major innovative countries in research total volume, contribution, and groundbreaking researchers by 2050.

The series of *Advances in China's Basic Research* and the series of *Reports of China's Basic Research* proposed and planned by the NSFC emerge against such a background. Featuring science, basics and advances, the two series are aimed at sharing innovative achievements, diffusing performances of basic research, and leading breakthroughs in key fields. They closely follow the frontiers of basic research developments in China and publish excellent innovation achievements funded by the NSFC. The series of *Advances in China's Basic Research* mainly presents the important original achievements of the programs funded by the NSFC and demonstrates the breakthroughs and forward guidance in key research fields; the series of *Reports of China's Basic Research* shows the core contents of the final reports of Major Programs and Major Research Plans funded by the NSFC to make a systematic summarization and give a strategic outlook on the achievements in the funding priorities of the NSFC. We hope not only to comprehensively and systematically introduce backgrounds, scientific significance, discipline layouts, frontier breakthroughs of the programs, and a strategic outlook for the subsequent research, but also to summarize innovative ideas, enhance multidisciplinary convergence, foster the continuous development of research in concerned fields, and promote original discoveries.

As Hsun Tzu remarked, "When earth piles up into a mountain, wind and rain will originate thereof. When waters accumulate into a deep pool, dragons will come to live in it." The series of *Advances in China's Basic Research* and *Reports of China's*

Basic Research are expected to become the "historical records" of China's basic research. They will provide researchers with abundant scientific research material and vitality of innovation, and will certainly play an active role in making China's basic research prosper and building China's strength in science and technology.

Wei Yang
Academician of the Chinese Academy of Sciences
Beijing, China

Preface

Rapid advances in science and technology are blurring the lines between traditional disciplines. Multidisciplinary cooperation centering on solving scientific problems has become a main driver to promote progress in various fields from basic research to translational research. Research in life science has developed from macro to micro aspects, from understanding basic rules with simple systems to exploring mutual relations with complex ones. It has become a focus of chemists to study scientific problems in the process of life using chemical concepts and methods. Since 2000, the research on such problems has gradually evolved from individual cooperation between chemists and biologists to further integration of chemistry and biology, making chemical biology emerge as a new field of life science. The studies of chemical biology on major problems in life science progress from static research to dynamic, from qualitative to quantitative, and from in vitro to in vivo. With the development of science, this discipline is thriving in China. From a strategic perspective of promoting leapfrog development in science and technology, the National Natural Science Foundation of China (NSFC) launched a Major Research Plan (MRP) during the 11th Five-Year Plan period. The Department of Chemical Sciences seized this unique opportunity of developing chemical biology in China, and refined scientific problems through discussions on the domestic and international progress of this field on many occasions from 2003 to 2006, such as the NSFC Shuang Qing Forum, Jiuhuashan Forum, and Xiangshan Science Conference of the Ministry of Science and Technology of China. The focus was finally defined as "cell—the basic unit in life activities—and its complex signal transduction network". The signal transduction determines the cell fate and directly affects physiological and pathological processes. Concerning the major scientific theme of cellular signal transduction, much could be done by chemical biology research. It was thus decided to investigate and monitor signal transduction pathways with small molecule probes to understand the structures and functions of this complicated life network. In 2007, the MRP of Investigations on Signal Transduction Processes Utilizing Small Chemical Probes was officially launched.

This MRP has attracted talents for promoting the development of chemical biology in China, and more importantly, fostered a research atmosphere that encourages multidisciplinary convergence. The 8 years, during which the plan was provided with continuous support, witnessed the rapid development of chemical biology in China. There were a growing number of chemists keeping up with the latest development and mastering experimental techniques of life science, and chemical laboratories developing into novel labs integrating chemistry and biology.

Great advances have been achieved through the 8-year efforts. Some newly-developed small chemical probes become general tools for signal transduction studies worldwide. In situ studies in living cell systems have been realized with the analysis methods for labeling and testing, bioorthogonal reactions, and the research methods of biophysical chemistry that have been developed during this period. New cellular signal transduction pathways and new mechanisms of physiological and pathological processes have been discovered, providing a basis for disease treatment. Biomacromolecules with the potential to be drug targets and promising lead compounds have been identified, creating drug research models based on chemical biology. In November 2016, at the conclusion evaluation conference organized by the NSFC, the series of final reports of this MRP were accepted by the expert team.

The MRP of Investigations on Signal Transduction Processes Utilizing Small Chemical Probes adopted a "5+3" funding mode. A top-level design based on extensive investigations was applied throughout the whole process from its establishment to implementation, and from project selection to result summarization, which helped to pool ideas and tackle key themes through joint efforts. The plan set up an Advisory Expert Group, Management Group, and Secretary Group. The Advisory Expert Group was composed of the Academician of the Chinese Academy of Sciences (CAS) Zhang Lihe, CAS Academician Lin Guoqiang, Academician of the Chinese Academy of Engineering Wang Hongyang, CAS Academician Chen Yeguang, CAS Academician Jiang Hualiang, Research Professor Wu Jiarui, Research Professor Chen Yi, and Prof. Xi Zhen. The Management Group consisted of Chen Yongjun, Feng Xuelian, Wu Lei, Du Canping, Zhuang Qiankun, Zheng Qiyu, and Zhang Yan. The Secretary Group, which assisted in compiling the final reports, was made up of Chen Peng, Liu Lei, Chen Xing, Huang Yanyi, Yang Caiguang, Wang Jiangyun, and other young scholars.

We are honored to be invited by the Center for Science Communication of the NSFC to compile the final reports of this MRP, which would be collected into the series of Reports on China's Basic Research. To present a full picture of the progress achieved by this MRP, we have also summarized the main findings of its nine integration projects. Research Professor Yang Caiguang was responsible for the compilation.

During the first 5 years of the plan, 132 fostering projects and 14 key projects were funded. Based on these research findings, nine integration projects with breakthrough potential were established. Hundreds of people have participated in this MRP and made contributions through painstaking efforts. In this regard, I hope to express my heartfelt gratitude to all participants.

Lihe Zhang
Academician of the Chinese Academy of Sciences
Beijing, China

Contents

Contributors

Yeguang Chen School of Life Sciences,Tsinghua University, Beijing, China

Yi Chen Key Laboratory of Analytical Chemistry for Living Biosystems Institute of Chemistry ,Chinese Academy of Sciences, Beijing, China

Hualiang Jiang Shanghai Institute of Materia Medica,Chinese Academy of Sciences, Shanghai, China

Guoqiang Lin Shanghai Institute of Organic Chemistry, Chinese Academy of Sciences, Shanghai, China

Hongyang Wang Shanghai Cancer Institute, Shanghai, China

Jiarui Wu University of Chinese Academy of Sciences, Beijing, China

Zhen Xi State Key Laboratory of Elemento-organic Chemistry(Nankai University), Tianjin, China

Chapter 1
Project Overview

1.1 Introduction

The Major Research Plan of Investigations on Signal Transduction Processes Utilizing Small Chemical Probes (hereinafter referred to as this MRP) was launched by the National Natural Science Foundation of China (NSFC) during the 11th Five-Year Plan period. The NSFC Department of Chemical Sciences had worked together with the Department of Life Sciences on conceiving and preparing this research initiative for years. After sufficient communication and discussions from chemists and biologists, this MRP of chemical biology was officially launched in 2007, and first released guidelines and accepted application in that January. Since then, it had officially released four guidelines for key and fostering projects (one in each year during 2007–2010), two for integration projects (one in each year during 2013–2014), and one Opinion on enhancing funding to the integration projects in several areas (2015). The 649 applications received were included in the fields covered by the 6 NSFC departments of Mathematical and Physical Sciences, Chemical Sciences, Life Sciences, Engineering and Material Sciences, Information Sciences, and Health Sciences, most of which fell into the range of departments of Chemical Sciences and Life Sciences (and the Department of Health Sciences established in 2009). Through screening via correspondence reviews and voting via panel committee meetings by peer experts including members of the Advisory Expert Group, a total of 160 projects were selected to be funded, consisting of 132 fostering projects, 14 key projects, 9 integration projects, and 5 strategic research projects. The applicant organizations of the project leaders were in 19 provinces, autonomous regions, and municipalities. There were 79 projects from applicant organizations under the Ministry of Education, 62 from those under the Chinese Academy of Sciences, and 19 from other organizations. The funding totaled RMB 200 million. All funded projects had been successfully completed by the end of 2015.

L. Zhang (ed.), *Investigations on Signal Transduction Processes Utilizing Small Chemical Probes*, Reports of China's Basic Research,
https://doi.org/10.1007/978-981-19-9585-9_1

1.1.1 Arrangement and Comprehensive Integration

This MRP was implemented in two phases. Phase 1, the fostering phase, was targeted on the core problems in the guidelines to collect projects from the public and encourage free exploration while fostering some projects in important areas. In this phase, a total of 606 proposals were received. After correspondence and panel committee reviews from experts, 124 projects were officially funded (including 110 fostering projects and 14 key projects). Before releasing, the four guides of the fostering phase were discussed by the Advisory Expert Group and were subject to supplements and adjustments concerning the main thread of the small molecule probe and the frontier development of related disciplines.

Phase 2, the integration phase, aimed to refine research focuses based on the research results in the fostering phase after the projects were evaluated by the Advisory Expert Group, to abstract major scientific themes, and thus increased funding provided to the identified projects. In total 32 well-completed projects with potential for breakthroughs were selected from the projects funded in Phase 1. According to the key supporting directions suggested in the Guide to Major Research Plan 2007, they were summarized to form 4 integration themes: (i) investigations on regulation mechanisms of glycolipid metabolism utilizing small molecule probes; (ii) investigations on the cellular signal transduction pathways induced by small chemical probes; (iii) investigations on molecular mechanisms of cell fate determination utilizing small molecule probes; and (iv) development of new analysis methods and techniques for studying signal transduction processes.

Additionally, the theme of "investigations on the nucleic acid related signaling pathways based on ligand regulation", which was naturally formed in the fostering phase, was abstracted and highlighted by the Advisory Expert Group. Given the international progress in the research on signal transduction processes utilizing small molecule probes, the promising theme of "regulation of some glycan-mediated cellular recognition processes", also formed in the fostering phase, was integrated. The guide on the 6 integration themes was released in 2012. After application, peer and committee reviews, 6 teams out of 10 were selected to be funded.

In 2013, more integration projects were funded according to the key research directions defined in the 2007 Guide. After screening among the well-completed projects in the three areas of probes and tools developments, detection methods, and pathway studies, the Advisory Expert Group invited applications from excellent young scientists in China to form 3 integration themes: (i) development and application of protein-based probes for signal transduction; (ii) new detection methods for studying signal transduction utilizing small molecule probes; and (iii) chemical biology studies on the regulation of cell plasticity in signal transduction. Three research teams with complementary advantages were built, highlighting the combined studies on methods, techniques, and specific biological processes, and focusing on breakthroughs in major scientific fields. In 2014, the research teams of the first integration projects were reviewed and 20 teams of well-completed subprojects were selected to receive renewed funding. Also, 2 research teams of projects closely related to this MRP were invited for proposals and received funding.

1.1.2 Interdisciplinary Cooperation

As a Major Research Plan in chemical biology, Investigations on Signal Transduction Processes Utilizing Small Chemical Probes had fully reflected the convergence of multiple disciplines. It has established interdisciplinary cooperation between biology and chemistry, including organic synthetic chemistry, natural product chemistry, medicinal chemistry, analytical chemistry, inorganic chemistry, and physical chemistry; and it also promoted the multidisciplinary collaboration among chemistry, medicine, material science, and biology. In the research area of signal transduction, the disciplines complemented each other. Cooperation in such way solved key scientific problems in biology, promoted frontier research in these fields, and attracted wide participation of expertise. The convergence of disciplines was embodied in four aspects.

(1) **Organic synthetic chemistry and cell biology**

Through utilization of the means of chemical synthesis and with a focus on key biomacromolecules in signal transduction pathways, the structural diversity of small molecule probes was put into great play to reveal the regulation rules and mechanisms in signal transduction pathways related to physiological activities. A series of diversified probes of high complexity were designed and synthesized to study the key molecular events in living cells, for example the important signal transduction processes mediated by JAK/STAT and Wnt. The methods for specific labeling and manipulation of biomacromolecules, including protein, nucleic acid, and carbohydrates, were highlighted since they could bring new insights into the regulation mechanisms of the involved signal transduction pathways.

(2) **Natural product chemistry and biochemistry**

The advantages of China's natural product resources were utilized to discover several "star molecules" that regulate cellular signal transduction pathways. Through the separation, purification and chemical synthesis of natural products, a series of natural products have been identified, which can enhance or inhibit the activities of several important signal transduction pathways.

(3) **Medicinal chemistry and medicine**

To realize the drug discovery model of "from functional genes to drugs", the combined studies on signal transduction processes and target discovery, validation of drug target functions along with compound screening were adopted as new research strategies to implement the fundamental research on new drug targets for several major diseases.

(4) **Analytical chemistry, physical chemistry, material science, and biology**

With the means of chemical analysis, new analysis methods and techniques were developed to collect information in signal transduction processes at multiple levels of molecules, cells, and living systems. With progress on frontier technologies such as microfluidics, single-molecule dynamic detection, and high-resolution imaging, their application in signal transduction processes were preliminarily explored.

Since it was officially launched, this MRP had always followed the general principle of "limited goals to enable steady supports, and integrated themes to realize leapfrog development", and carried out innovative research concerning the frontiers of the interdisciplinary fields of chemistry and life science. The interdisciplinarity was actively promoted through the enhancement of the top-down design and refinement of scientific goals. It had significantly facilitated the convergence of chemistry and multiple disciplines, such as biology and material science, and multidirectional extensions of the research areas of chemical biology to greatly promote the development of this discipline in China. It had also fostered a team of talents in chemical biology and made China's research in this field gradually give its own voice in the world. In 2008, the Advisory Expert Group and the Management Group were invited to publish an article in *Nature Chemical Biology* to introduce this MRP and the research of chemical biology in China.

1.2 Research Overview

1.2.1 Overall Scientific Objectives

This MRP aimed to utilize the advantages of the convergence of chemistry and multiple disciplines such as life science and medicine to make breakthroughs on the traditional research methods in single disciplines of chemistry and life science. With small molecule compounds as probes, it endeavored to develop new methods and technologies, study key molecular events in signal transduction processes in living systems, uncover the mechanisms of regulatory signaling in cells, and prove clues for the discovery of biomarkers, drug targets, and lead compounds for diagnosing and treating major diseases. It also sought to promote the convergence of research in chemistry and life science to form growing points for new disciplines.

1.2.2 Key Scientific Issues

This MRP aimed to explore the following 4 key scientific issues.

(1) **Generation of small molecule probes for studying signal transduction**

Bioactive small molecules of diversified structures were important starting scaffolds and functional probes for studying signal transduction mechanisms, and are key sources of lead compounds for drug discovery. Although China had established a solid foundation for identification and synthesis of natural products, these strengths had not translated into enhanced discovery of chemical probes for studying biological functions, especially those small molecule probes with important values, owing to the lack of collaboration between chemists and biologists in China. This MRP focused on studying the regulation mechanisms in signal transduction processes and related cellular physiological functions, developing new methods and technologies to obtain active small molecule compounds, and improving the capabilities for creating small molecule probes. Additionally, these projects have accumulated chemicals and gradually established a shared library of active small molecules and probes.

(2) **Development of detecting methods for gathering information of biomacromolecule activities and interactions in living systems intervened by small molecule probes**

The signal transduction processes are involved in many interactions between multiple molecules, including interactions between biomacromolecules in cells (such as protein-protein interactions, and protein-nucleic acid interactions), and cellular recognition of molecules, and interactions between biomacromolecules and small molecules. These interactions formed complex networks to perform different functions and actions. Obtaining the information of the intermolecular interactions in living systems was a pivotal step towards elucidating mechanisms of signal transduction.

(3) **Investigations on mechanisms of signal transduction based on small molecule probes**

The signal transduction is involved in transferring the information generated at the cell membranes to intracellular effector systems such as nuclei, as well as transmitting signals between cells. Those processes were usually studied based on interactions of multiple molecules such as proteins, which were highly reversible and transient processes. Applying chemical biology methods to the study of signal transduction showed some definite advantages. For example, the research could be steadily performed in controllable and reversible manners. Compounds could be added or removed at any time to initiate or terminate a specific reaction in cells. Most small molecule compounds acted on protein targets quickly and could be tested in a real-time manner. Moreover, the dynamic processes of the corresponding targets could be analyzed by changing compound concentrations. Another advantage was that the same compounds could be widely used to affect a specific process or function in different organisms. This MRP applied small molecule probes in chemical biology research, which focused on important molecular events in signal transduction processes in living systems to uncover the regulatory mechanisms of cell signaling.

(4) **Discovery of disease biomarkers, drug targets, and lead compounds based on signal transduction processes**

Based on the results from the themes mentioned above, further investigations were conducted on functions and interaction networks of important proteins, therefore, to discover biomarkers, drug targets, and lead compounds related to diseases and lay a foundation for disease prevention, diagnosis, and treatment.

1.3 Significant Progress

This MRP had given full play to the advantages of multidisciplinary cooperation of chemistry and life science and utilized small molecule probes as main tools to further study key molecular events and their mechanisms in signal transduction processes. Particularly, breakthroughs had been made in the research on cell fate regulation, glycolipid metabolism, pluripotency maintenance and reprogramming, as well as a series of classical signal transduction pathways. It has revealed the molecular mechanisms of physiological processes, such as cell proliferation, differentiation, apoptosis, migration, and reprogramming, analyzed the relations between cellular signal transduction pathways and epigenetic regulation, and suggested the drug targets specifically intervening the corresponding pathological processes.

Through over 8 years' efforts, this MRP had successfully fulfilled the intended tasks and achieved the overall scientific goals. Given the evaluations of its implementation by the Advisory Expert Group, this MRP had made major progress in the proposed 4 key scientific issues.

1.3.1 Development of Small Molecule Probes (>300 Types) Based on Natural Products and Chemical and Biological Synthesis, and Discovery of Multiple "Star Molecules"

This MRP studied the cellular signal transduction pathways in important biological processes, and systematically obtained a set of novel small molecular probes with independent intellectual property rights. More than 300 types of small molecule probes were developed based on natural products, chemical and biological synthesis, and multiple "star molecules" were obtained, which provided molecular tools for revealing signal transduction processes, discovering new signal transduction pathways, and understanding related physiological and pathological mechanisms

(Fig. 1.1). Organic synthesis technologies were developed for building diversity-oriented compound libraries, which were utilized as key methods to synthesize a group of small molecule probes with unique structures. Biosynthesis methods were advanced for efficiently producing bioactive small molecules, which were used as effective tools for obtaining the probes to target biomacromolecules such as proteins and nucleic acids. This MRP also exploited the unique and abundant resources of natural products in China and built featured libraries of active molecules with the techniques of chemical and biological synthesis as well as separation of natural products. Additionally, it developed efficient chemical synthesis methods for building combinatorial libraries of natural products, complex natural glycoconjugates and oligosaccharides, cyclopeptides and polypeptides with different modified groups. It also developed biological synthesis methods for obtaining active molecules. These efforts have provided novel molecular probes for a series of research on cellular signal transduction pathways. Utilizing these probes, the research groups elucidated biological processes such as autophagy, apoptosis, and directed migration of immune cells, and revealed the molecular mechanisms of signal transduction pathways including TGF-, G protein-coupled receptor, Wnt, TNF-α, and NF-κB. They identified active ingredients of natural products and obtained the binding targets as well, which would promote the drug discovery for treating diseases such as leukemia, lysosomal storage, cerebral stroke, sepsis, obesity, and Alzheimer's disease.

1.3.2 Development of Special Techniques and Frontier Analysis Methods of Chemical Biology Including the Bioorthogonal Reaction and Single Molecule/Cell Detection

Facing the technical challenges in exploring the cellular signal transduction processes, this MRP brought classic strategies of chemical biology, such as the bioorthogonal reaction, into the research, emphasized their optimization and application in living systems, and highlighted their advantages of high specificity, sensitivity, and resolution. Major breakthroughs had been achieved in chemical labeling, probe development, imaging method, omics technology, separation method, microfluidic method, sensor technology, and other technologies.

Based on the research contents and needs, this MRP created a series of distinctive chemical methods for labeling, activating, and manipulating biomacromolecules, and achieved the active manipulation of accurate labeling, highly sensitive imaging, and targeted regulating in high temporal-spatial resolution on proteins, glycans, and nucleic acids in living cells and animals. It invented key analysis technologies and methods that were sensitive and practical, especially those analyzing organisms with interconnectivity, full scale, and single molecule/ cell resolution. The measurement methods were transformed from the previous ones that extremely depended on the ensemble average to those of high accuracy, dynamics and sensitivity that targeted

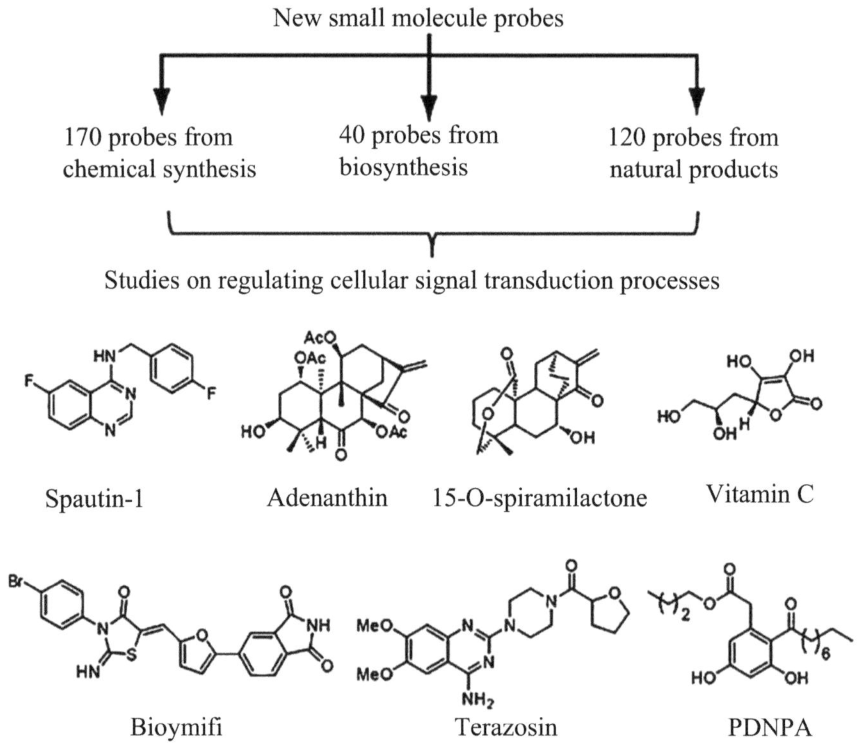

Fig. 1.1 Representative probe molecules

on individuals, especially single molecules or cells. Vital biological information was obtained at the levels of molecules, cells or living animals. This MRP also developed multiple optical probes for detecting important bioactive molecules, such as metal ions, free radicals, and reactive oxygen species, in living cells in a real-time manner, and *in situ* detection sensors for glycosyl sites and glycans on cell surfaces as well. It established new methods for detecting proteins and nucleic acids using chemical antibody-aptamer recognition, and developed mass spectrometry- and optical detection-based methods for profiling the structural complex of proteins bound with drugs and ligands and revealing recognition mechanisms.

1.3.3 Elucidation of Key Signal Transduction Processes Including Cell Reprogramming, Glycolipid Metabolism, Cell Apoptosis, and Wnt Signaling Pathways

In research on signal transduction processes, small molecule probes were easily controlled for regulation, quickly functioned and always produced apparent quantitative relationships. These distinctive advantages were especially fit for studying the signal transduction processes that were highly dynamic and transient. Therefore, this MRP brought together expertise in the interface research of chemistry and biology. They focused on important nodes of cell fate regulation and key molecular events in signal transduction processes by utilizing active small molecule compounds to explore the regulation rules and the molecular mechanisms of cell fate determination which is either differentiation or death. It had mastered the molecular patterns of cell reprogramming and plasticity regulation, discovered new mechanisms regulating glycolipid metabolism, and further understood the working modes of several classic signaling pathways. For instance, this MRP revealed new pathways and related regulatory targets of processes including leukemia cell differentiation, autophagy, and tumor apoptosis, discovered regulatory mechanisms for negative feedback in cholesterol and insulin metabolism, improved the efficiency of cell reprogramming, developed efficient inhibitors targeting several signaling pathways, and analyzed the dynamics and gating mechanisms of acid-sensing ion channels.

1.3.4 Discovery of New Targets and Lead Compounds for In-fectious, Metabolic, and Inflammatory Diseases, and Tumors

On the basis of the development and application of novel small molecule probes for studying important cellular signal transduction pathways and major biological events in signal transduction, this MRP also focused on solving the basic problems of drug discovery for main diseases, as well as fundamental topics of drug target discovery and validation with new concepts. It had achieved significant progress in applying small molecule probes and new targets to early development of first-in-class drugs. In important fields of drug discovery, including tumors, immunological disorders, neurodegenerative diseases, metabolic dysfunction, and infectious diseases, it emphasized the studies on novel protein targets in the important pathways, including metal-ion transport, ubiquitination signal transduction and regulation, two-component sensing systems, glycolipid metabolism, and cell differentiation, and completed effective functional research and verification of target druggability. This MRP also obtained small-molecule lead compounds with novel structures derived from natural products or lab-made compounds, identified new sites and functions of known targets, and licensed the promising lead compounds to pharmaceutical companies. These studies had accelerated the identification and validation of new targets. In addition, it realized the successful formation of new patterns in drug discovery—transformation from the classic patterns based on experiences and pharmacological models to new patterns of precise targeted drug discovery based on molecular targets and disease genotypes. These achievements had fully shown the core position of chemical biology in innovative drug discovery and the important contributions it made.

After the completion of this MRP, the development trends of the related important fields are compared in Table 1.1.

Table 1.1 Comparison of development trends

Field	International background	At the beginning of this MRP	At the end of this MRP
Chemical probe	Thriving of "small-molecule intervention"	Narrow in categories, and short on quantity	Full range, and abundant in quantity Discovery of multiple star probes were obtained
Special techniques of chemical biology	Special techniques of chemical biology were gradually developed	Almost empty	Development of various mainstream research techniques
Research on mechanisms of signal transduction	Research on signal transduction utilizing small molecules and traditional methods complemented each other effectively	Lack of systematic work	Analysis of a series of key signal transduction pathways
Drug target and chemical intervention	A new pattern of drug development based on chemical biology	Lack of research led by chemical biology	Discovery of targets and compounds led by chemical biology

Chapter 2
Domestic and Foreign Research

The beginning of the twenty-first century witnessed unprecedented development of chemical biology as an emerging discipline. More and more chemists in traditional disciplines are collaborating with biologists in the field of chemical biology. With a focus on solving scientific problems and developing research tools, such research has significantly promoted our understanding of complex living systems. They utilize what they are good at—chemical synthesis strategies, chemical reaction techniques, and small molecule probes—to label, modify, and chemically intervene in biomacromolecules and the living systems. They have also developed next-generation analysis and detection technologies that are more sensitive and accurate for systematic observation, detection, and manipulation of underexplored living systems. The two complementary kinds of research tools have jointly promoted and enriched our research strategies of life activities. The research in chemical biology mainly focus on revealing mechanisms of physiological processes on the molecular level, with exogenous chemical reactions or molecules to regulate and intervene in living systems, which paves the way for drug development in the future. This chapter describes the domestic and international progress in chemical biology regarding the development of special tools, such as small molecule probes and biorthogonal reactions, advances in biological detection techniques, and biology-orientated problem probing.

2.1 Thriving of "Small-Molecule Intervention"

Bioactive small molecules of diversified structures are important probes for studying signal transduction mechanisms, and key sources for identifying lead compounds for drug discovery. The screening and design of bioactive small molecule probes are among the most prevalent chemical biology tools and prove successful in extensive fields from fundamental research to clinical translation. During the process, the idea

L. Zhang (ed.), *Investigations on Signal Transduction Processes Utilizing Small Chemical Probes*, Reports of China's Basic Research,
https://doi.org/10.1007/978-981-19-9585-9_2

of the chemical probe has lodged itself in the community and become one of the active topics with most intensive cooperation between chemists and biologists.

2.1.1 Diversity-Oriented Functional Compound Library

As an increasing number of active small molecules are used for developing probes and analyzing signal transduction pathways, scientists are no longer satisfied with drugs and bioactive natural products. They strive to find small molecule probes with new functions in chemical biology. With the development of this discipline, the libraries of small molecule compounds with diversified structures are frequently screened by medicinal chemists to identify lead compounds for drug discovery, and utilized by chemical biologists to develop small molecule probes, with more attention paid to the various synthesis, such as the diversity-oriented synthesis (DOS). Additionally, unlike medicinal chemists who need the newly developed small molecule probes to be druggable as lead compounds, chemical biologists have higher requirements for the biological activity and specificity to probe the biological mechanisms of signal transduction pathway regulation.

2.1.2 Early Exploration of Drug Discovery

The model of ligand-directed drug discovery that is mixed with chemical biology as a new strategy has become popular for drug target validation and lead compound identification in many drug discovery institutes. For instance, the Bump-and-Hole strategy has opened up a new way to investigate the activity of specific kinase with small molecule compounds, which has been extended to research of other proteins. This idea is not only widely employed in fundamental research but also successfully used in drug development. Several new candidates have now entered clinical trials for drug discovery. The functional genomics research based on the libraries of small molecule compounds has been a big success. Currently, various drug precursors in clinical assessment for treating cancers and immune or infectious diseases have been developed through such methods. Small molecules promoting or suppressing the growth, differentiation, or regeneration of stem cells have been screened and identified from functional small molecule libraries, which indicates promising applications of small molecule compounds in regenerative medicine. China enjoys abundant natural resources of land and sea as well as rich resources of Chinese medicinal herbs, and has a large number of scientists active in natural organic chemistry, synthetic chemistry, and analytical chemistry. During recent years, they have partnered with biologists to analyze various molecular pathways of signal transduction with structurally diverse natural products and synthetic compounds as well as evolving analysis methods. Significant achievements have been made, and some findings are expected to achieve clinical translation.

2.2 Extensive Introduction of Biomolecular "Un-natural Analogues"

Nature has developed two types, template-based and non-template-based machines, for biosynthesis of biomolecules and complexes with versatile functions. The template-based machine is composed of ribosomes, DNA/RNA synthases, and polyketide synthases, and the non-template-based machine includes metabolic enzymes. In addition, the modification, transformation, and evolution of biomolecules and their functions are typically achieved using the genetic methods of mutation, gene editing, recombination, and amplification. Accumulated evidence has implied that chemists have mastered the ability of utilizing such biosynthetic machines to create molecules with novel functions. These efforts have focused on the artificial transformation of building blocks to the biomolecules and provided enhanced or even new functions built on them. This idea has been widely adopted in the synthesis of polyketides, ribosomal and non-ribosomal polypeptide antibiotics, glycopeptides, secondary metabolites, and bioenergy. Chemical biologists pay more attention to changing the structures and functions of biomacromolecules using this technique and extend it to all kinds of them. Notably, the concept of "unnatural analogues", such as the unnatural forms of amino acids, sugars, nucleic bases, and lipids, is a major research topic in chemical biology. They have been created due to the high integration of exogenous chemistry and biosynthetic machines. The introduction of "unnatural analogues" has not only enriched the tools in understand-enabled-manipulation of physiological processes, but also distinguished chemical biology from other traditional disciplines in terms of concepts and thinking.

2.3 Extension of Exogenous Chemical Reactions to Living Systems

During recent years, to remain biomacromolecules and complex living systems intact, chemical biologists have developed specific tools for labeling, manipulating, and studying the complicated biological machines in the hope of achieving direct manipulation in in situ studies. It is against this backdrop that the concept of biorthogonal reactions has been put forward.

This term refers to any chemical reaction inside living cells or organisms without interfering with endogenous biomolecules. The introduction of biorthogonal reactions is a breakthrough in chemical biology, which remains one of the hot research topics. Before 2000, small molecules were adopted to study living systems by chemists who were actively involved in life science research. Since the beginning of the twenty-first century, advances in biorthogonal reactions have significantly enhanced the capabilities of labeling, tracking, and monitoring biomacromolecules in vivo. This technology and the biorthogonal chemistry have grown as general tools for studying and analyzing molecular targets in living systems. Chemistry has been more widely applied in research of life science, indicating its essential role in exploring the complicated living systems. Additionally, the thriving of biorthogonal reactions has also substantially promoted the development of reagents and methods to catch and fish target proteins, as well as to monitor a molecule or single cell. They all provide powerful tools for research in chemical biology.

2.4 Efficient Chemical Synthesis of Biomacromolecules

In recent years, scientists have developed efficient chemical methods for the synthesis of biomacromolecules, including cyclopeptides and polypeptides with different modified groups, proteins, natural glycoconjugates and oligosaccharides, nucleic acids, and their analogues. The *in-vitro* screening and *in-vivo* evolution of functional nucleic acids (aptamers and nucleic acid enzymes), polypeptides, and proteins have significantly extended the channels for the manual acquisition and functional optimization of these biomacromolecules. As the chemical synthesis capability grows and the understanding of living systems deepens, chemical biologists are increasingly interested in chemically synthesizing the bioactive macromolecules, such as proteins consisting of over 100 amino acids, proteins with specific chemical modifications, complicated glycoconjugates, and glycoproteins. For instance, the rapid development of native chemical ligation (NCL) has promoted the studies of protein structures and mechanisms. The structures of synthetic erythropoiesis proteins (SEPs), plectasins, ester insulins, and vascular endothelial growth factors (VEGFs) and the ligand-bound complexes, have been analyzed through chemical synthesis and racemic crystallography, which has far-reaching significance on protein research and medical applications. Many scientists have also completed competitive work in this field in China.

2.5 Rapid Advances in Biological Detection Technologies

Advanced analysis techniques are vital to reveal the mystery of biological activities in living systems, and they are core drivers of life science research. With rapid development of more sensitive, accurate, and spatiotemporal specific analytical methods, biologists could utilize them to detect and explore the living systems at subcellular levels and molecular precise.

2.5.1 Bioimaging

Bioimaging is undoubtedly one of the representative technologies with far-reaching influences. As a powerful tool to understand and explain many life phenomena, bioimaging enables clear manifestations of mysterious living systems, while the development of imaging probes is a crucial element. These tools are composed of small-molecule dyes and fluorescent protein-based genetic probes. Their applications range from exploring structures, spatial distributions, and dynamic changes of target proteins to directly monitoring these proteins in internal cell structures, subcellular organelles, and living animals. Chemists have contributed greatly to the development and utilization of bioimaging probes. The super-resolution imaging is one of the fastest-growing and most eye-catching optical imaging technologies. It has improved the microscope resolution from hundred-nanometer to ten-nanometer, a significant increase in the precision of biological imaging. Chemical biology has contributed significantly to the development and optimization of fluorescent dyes, labeling techniques, and data processing methods for super-resolution imaging, as well as the promotion of its applications in the fields of life science.

2.5.2 Chemical Proteomics

Based on mass spectrum and proteomics technologies, chemical biologists pay special attention to the activity-oriented proteomic research and develop methods for specific capture, enrichment, and identification of the protein components with desired activities in the whole proteome. The activity-based protein profiling (ABPP) can be applied together with small molecule probes to detect the activities of a specific enzyme family in living cells, provide detailed functional analyses of many important target proteins, and perform quantitative profiling of chemical modifications on proteins with interest. Chemical proteomics has currently become a crucial promising part of chemical biology.

2.5.3 Chemical Modifications of Nucleic Acids and Next-Generation Sequencing Technologies

Research on the important biological functions of chemical modifications of biomacromolecules, including methylation, acetylation, and ubiquitination, and their relations with human diseases, is one of the most active frontiers in the convergence of chemistry and biology and in life science. The development of research tools and strategies based on chemistry has dramatically promoted advances in this field. For instance, the combination of highly specific enzymatic and chemical reactions has enabled the selective and specific labeling of new chemical modifications of DNA, such as 5-hydroxymethylcytosine. In this way, such modifications can be successfully detected with their specific output signals in the high throughput sequencing. In addition, the advances in chemical labeling and reactions are indispensable to the development of new gene sequencing technologies, such as the single-cell whole-genome sequencing (scWGS) and the deep sequencing at single-base resolution.

2.5.4 Single Molecule Detection and Single Cell Analysis

In recent years, the single molecule detection has made great progress, making it a powerful tool for studying biomacromolecules in living systems at the level of a single molecule. Meanwhile, the single cell analysis is developing fast, which enables the analysis of the chemical components in a single cell in a quantitative and real-time manner. Chemical biologists are endeavoring to realize the real-time in situ measurement, even manipulation, of dynamic changes and chemical modifications of a single biomacromolecule, as well as the involved interactions and biochemical reactions.

2.6 Research on Biology-Directed Scientific Problems

The development of chemical probes, orthogonal reactions, analysis technologies, and their continuous extensions into life science is me of the core drivers of the emergence of chemical biology. Another critical driver originates from life science itself. Since the beginning of the twenty-first century, lifeomics research, represented by the Human Genome Project, has blossomed, accumulating a large number of data for biological research and raising many new biological problems. However, it is difficult to study them only with the classic genetic central dogma and traditional biological technologies and methods. To cope with this challenge, it is particularly important to develop efficient and universal research tools. The research technologies and interventions based on chemistry have stood out during the process and become critical methods and valuable tools in biological research. They have played a decisive role in biological research and function discovery, such as transduction signaling in cells. The demands for answering specific scientific problems in life science have been closely integrated into both domestic and international research. An increasing number of chemists switch their research ideas from pure chemistry to interdisciplinary fields, and they come to pay more attention to developing methodologies useful in the special environments of organisms. More chemical laboratories have established systems and facilities for biological research, and chemical biology laboratories emerge and grow. In summary, a new research culture, which targets scientific problems key in life science and emphasizes multidisciplinary cooperation and convergence, is being formed in chemical biology. This research model of close cooperation and high integration of chemistry and biology has extended to many fields of life science.

The overview of both domestic and international progress in chemical biology shows a strong trend towards further integration of developing chemical tools and answering a specific biology-directed question, which is an innovative source and a sustainable driver of this discipline. Its future development and perspective are expected to be as follows.

The development of small molecule probes and chemical tools will gradually transit from the in vitro studies to the in vivo investigations, from biomacromolecules and cells to more complicated systems such as living tissues and animals. More emphasis will be placed on the development of methodologies in the particular environments of living systems. The general application of the developed tools and their irreplaceability in biological research will be further highlighted.

The studies using biological detection technologies will be steadily transformed from static to dynamic detection, from qualitative to quantitative measurement, from in vitro to in situ exploration, to reveal the molecular mechanisms of life science more precisely and sensitively. Especially, the frontiers of biology are fast-evolving and highly timely. Highly relevant detection tools must be efficiently designed and developed to satisfy the rapidly changing demands for probing biological problems.

The convergence of chemistry and biology will deepen, and the research on biological problems will go further. The pattern that biologists raise questions, develop tools and study them with the assistance of chemists will be gradually changed. Instead, it will be chemical biology that truly promotes the process of putting forward and answering scientific questions in life science. During the cooperation between chemistry and biology, new biological questions will be brought out, and chemical tools and detection methods will be developed correspondingly to reveal mechanisms of biological science and human health further. During the process of solving problems in biology, chemistry itself will also grow dramatically.

More attention will be paid to integrating theory with practice, promoting translational applications, and realizing the whole chain development of upstream and downstream connection in order to substantively contribute to social development and human health. Take the early stage in discovery and exploration of the novel drugs as an example. Chemical biology has profoundly affected innovative activity and changed the established mode of drug discovery. The combined strategy of hits screening and drug targets validation in the beginning of drug discovery is gradually maturing. The research model of "from functional genes to drugs" will be widely applied in the whole chain of drug discovery such as target discovery, functional validation, hits identification, and optimization of lead compounds.

Chapter 3
Major Research Achievements

The representative achievements of this MRP are summarized in the chapter.

3.1 Specific Regulation Technologies Targeting Proteins in Cellular Signal Transduction

As one of the largest groups of biomacromolecules in organisms, protein is a main performer of biological functions and plays a main role in various life activities. The protein manipulating tools applicable to living systems can enable in situ studies on the structures and functions of target proteins, making them one of research hotspots in chemical biology. Particularly, compared with traditional chemical probes and techniques for activity inhibition, the techniques that can activate proteins with highly spatiotemporal specificity have great advantages. They are key techniques for studying the cellular life processes involving proteins, especially signal transduction processes. Although ultraviolet light has been used for specific regulation of protein activity, its features of poor penetration and high cytotoxicity significantly limit its application in living systems. The integration project of Development of Protein-Based Probes and Their Applications in the Studies on Signal Transduction has developed chemical decaging, two-photon and infrared light regulation, as well as induction strategies of targeted small-molecule probes, enabled specific activation of proteins in cell, and then utilized such new technologies and methods to study the molecular mechanisms of cellular signal transduction processes.

L. Zhang (ed.), *Investigations on Signal Transduction Processes Utilizing Small Chemical Probes*, Reports of China's Basic Research,
https://doi.org/10.1007/978-981-19-9585-9_3

3.1.1 *New Technology Targeting Protein Kinase—Chemical Decaging*

The phosphorylation mediated by protein kinases is one of the core processes in cellular signal transduction pathways. It has an important impact on most physiological processes. The signaling pathways involving kinases are also related to multiple diseases, including cancer, metabolic and immune diseases. However, the molecular mechanisms of many kinases in normal physiological and pathological conditions remain unclear. Utilization of small molecule activators can help to get new information from the studies on kinases in signal transduction. To make the small molecule activation strategy generally applicable to kinase research, the research group of CHEN Peng in Peking University combined the site-specific incorporation of unnatural amino acids and bioorthogonal decaging with chemcials, and reported a new strategy for rationally designing small-molecule enzyme activators [1]. A crucial lysine residue (a conserved site) on a kinase could be caged with the genetic code expansion technique to mask its activity, and the decaging reaction was adopted to restore its activity from the inhibitory state, as shown in Fig. 3.1. To implement this strategy, they developed a palladium-mediated depropargylation reaction that was successfully applied to proteins within living cells. Then, they utilized the bioorthogonal elimination reaction to study the influence of bacterial effector protein on the kinase ERK of host cells. The effector protein OspF was activated within living cells, and it was observed to irreversibly dephosphorylate the phosphorylated ERK in the nucleus, which led to the ERK protein exported from the nucleus to the cytoplasm. Additionally, they cooperated with the group of CHEN Xing, also from Peking University, to apply this reaction to the chemical regulation of cell-surface glycans to manipulate cell-surface charge [2].

CHEN Peng's group designed a special inverse electron-demand Diels–Alder (invDA) reaction to mediate chemical decaging. In this reaction, the carbon–oxygen bond at the allylic position broke after a cycloaddition reaction, and the trans-cyclooctene as a caged group was eliminated. It was been verified on the firefly luciferase within living cells [3]. Additionally, it also inherited the features of the invDA reaction that was efficient, rapid, and bioorthogonal. It could activate more

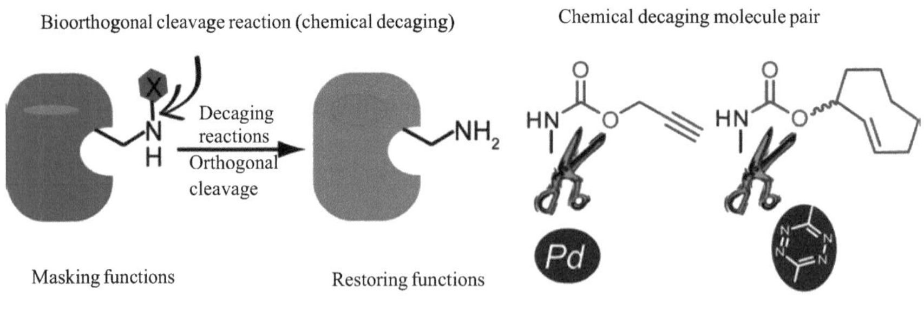

Fig. 3.1 Chemical decaging for activating proteins in vivo

than 80% caged protein kinases within 15 min. This breakthrough is of great significance, driving "the rational design of protein activation" from the proof of concept to applications. The technology has been successfully used to specifically activate important kinases, such as MEK-1, FAK, and protooncogene Src kinase, which may provide an efficient tool for studying molecular mechanisms of kinase-related diseases [4]. In their invited "Perspective" article in *Nature Chemical Biology*, they summarized the chemical decaging technique based on bioorthogonal cleavage reactions as a thriving area of chemical research, and proposed the future directions [5].

3.1.2 Light-Controlled Protein Probes Targeting Immune Cell Activation

In recent years, light-controlled proteins have become a powerful tool for studying the spatiotemporal regulation of cellular signal transduction. Compared with ultraviolet-activated probes, two-photon-activatable probes can significantly reduce cytotoxity, and thus have a broad application prospect. The research group of LIU Lei from Tsinghua University utilized protein chemical synthesis as a core technique, and developed light-controlled protein probes targeting immune cells. With these probes, they studied the directional migration of immune cells under precise spatiotemporal stimulation and the activation mechanism. Through the development of two-photon-controlled chemokine probes, they and their collaborators explored the mechanism of directional migration of immune cells in living tissues. They designed and synthesized the first two-photon-activatable chemokine probe hCCL5**. With its high spatiotemporal resolution, they concluded at the single-cell level that "**PI3K is not responsible for perceiving the direction of T cells, but is nonetheless required for persistent migration**" [6]. Additionally, utilizing the good tissue penetration of the two photon technique, they and their collaborators realized the directed migration of immune cells in living tissues (ears and lymph nodes of mice). This probe is an ideal molecular tool for understanding and potentially manipulating cell positioning and related cell biology. Based on the developed light-controlled protein antigens, they revealed the early signal transduction dynamics of B cell activation. HEL-K96NPE, the first light-controlled protein antigen for B cells, was developed with light-controlled amino acid screening and protein chemical synthesis. They found that for hen egg lysozyme (HEL) and its antibody HyHEL-10 (or B cells from MD4 transgenic mice expressing HyHEL-10), masking a single site (Lys96 of HEL) was sufficient to block the interaction as high as 20 pM with a large interaction surface (1800 $Å^2$). Combined with the high-speed high-resolution live cell imaging, HEL-K96NPE was used to monitor the early formation of B cell synapse and the periodic responses of calcium influx in a real time manner. This probe is a promising tool for in-depth studies on the early signal transduction dynamics of B cell activation.

3.1.3 Near-Infrared Light-Controlled Activation Targeting Cell Surface Receptors

A cell uses its surface receptors to identify the ligands in the environment and responds accordingly, such as directional migration, activation, proliferation, and differentiation. The light-controlled protein activation has been widely used in studies on the mechanisms of signal transduction mediated by interactions between the cell surface receptors and their ligands. However, many of these light-sensitive proteins are activated by ultraviolet or visible light, which has limited their tissue penetration in living systems. Near-infrared light is ideal to regulate signal transduction due to its advantages such as deep tissue penetration and small phototoxicity to cells. The group led by CHEN Xing from Peking University reported a single-walled carbon nanotube (SWCNT)-assisted near-infrared protein activation technique to regulate the transforming growth factor β (TGF-β) signal transduction. The TGF-β signaling pathway is crucial in the regulation of the embryonic development through activation at specific time points and sites. In living systems, TGF-β forms the small latent complex (SLC) with latency-associated peptide (LAP) to inhibit its activity, and the release of TGF-β enabled its activation to realize the spatiotemporal regulation. The SLC was conjugated onto SWCNTs, and TGF-β signaling pathways were activated due to the local heating generated by the characteristic near-infrared light absorption of carbon nanotubes [7]. They verified this strategy at living cell level for the first time. Irradiation by near-infrared light resulted in successfully modulating the cell differentiation, migration, etc. mediated by TGF-β signal transduction. This method was also applied to the near-infrared light activation of TGF-β signal transduction in mice.

3.1.4 Probes for the Generation and Degradation of Target Proteins

Chemical control of protein contents in living cells is another powerful method for regulating protein functions and activities. The research group of YANG Yi from East China University of Science and Technology used natural photosensitive probes to establish a convenient Light On photo-switchable gene expression system which enabled the precise spatiotemporal control of protein expressions in living cells or animals. With this system, they successfully controlled the insulin generation and blood glucose level in diabetic mice [8]. The Light On system enabled efficient control of protein contents in complex biological systems, and provided a new method for time- or dose-adjustable gene therapies. All of its components are flexible and adjustable, which makes it applicable to various research. This achievement has applied for both Chinese and international PCT invention patents. Additionally, they combined chemical and light regulation techniques to enable the control of intracellular gene expression by small molecules and light, which could reduce the background of the system under non-induced conditions and improve its sensitivity [9]. According to the cell physiological characteristics of model organisms such as zebrafish, drosophila, nematode and yeast, the expression system that could be induced efficiently in a variety of non-mammalian prokaryotic cells was optimized. Efficient light-controlled repressor proteins and reverse transcription terminators were obtained with the switch efficiency increasing over 10,000-fold, which was the most stringent known gene expression system. A light-controlled protein degrader was acquired to control the degradation of target proteins in cells by light, with the switch efficiency increasing over 20-fold, making it of practical value. The research group of LI Yanmei from Tsinghua University utilized the controllable protein degradation strategy to reduce the intracellular target protein level, in order to reduce its activity. The level of Tau, an important nonenzymatic protein in Alzheimer's disease, increased abnormally in patients. Therefore, they used a "recognizing and cleaving" strategy to regulate the intracellular level of Tau. The segment *VQIVYK* crucial for Tau aggregation was selected to be the specific recognition moiety, and *Cyclen* chelating Cu^{2+} as a cleavage motif which functioned as an active center of metalloenzyme to hydrolyze the protein. This active molecule could inhibit Tau aggregation in vitro. The group further added a cell-penetrating peptide to the end of the molecule, so that it could enter into cells to degrade the intracellular Tau protein [10]. They also screened efficient bifunctional molecules to target the proteins of interest, and utilized ubiquitination systems to achieve the small-molecule induced ubiquitination of specific proteins to degrade them. In research on cellular signal transduction processes, it is necessary to down-regulate the level of a specific protein to explore the impact of its absence on signal transduction. This technique is an important tool to degrade misfolded proteins and study the related regulations.

3.1.5 Ultra-Bright Photoactivatable Fluorescent Proteins

Super-resolution imaging techniques can be used for observing dynamic processes, such as protein conformational changes, protein interactions, and spatiotemporal regulations of signal transduction, making them significant approaches to high-resolution monitoring and analysis of cellular signal transduction networks. The gene-encoded green fluorescent protein and its variants are one of the most powerful tools for protein labeling and imaging. The current challenge is to develop fluorescent protein variants that are brighter with higher conversion rates, in order to achieve super-resolution imaging. **The performance of super-resolution imaging relies heavily on the characteristics of photoactivatable fluorescent proteins (PAFPs)**, including the key properties of **size, brightness, maturation rate, oligomeric nature, pH stability and photon budget after switching. Also, although often neglected, the label density limits the effective spatial resolution in all super-resolution fluorescence microscopy**. The research group of XU Tao from the Institute of Biophysics, Chinese Academy of Sciences (CAS), designed a novel monomer photoactivatable fluorescent protein that could be used in super-resolution

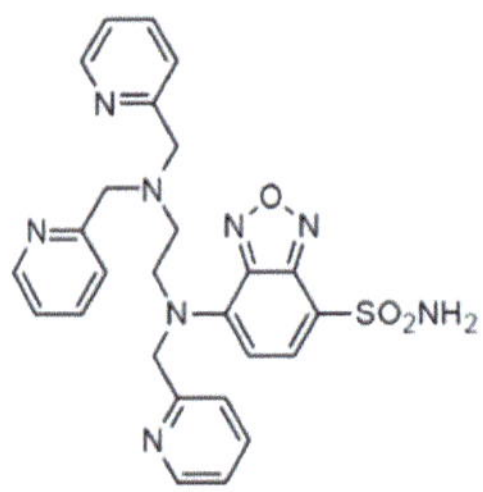

Novel ratiometric fluorescent probe for zinc ion

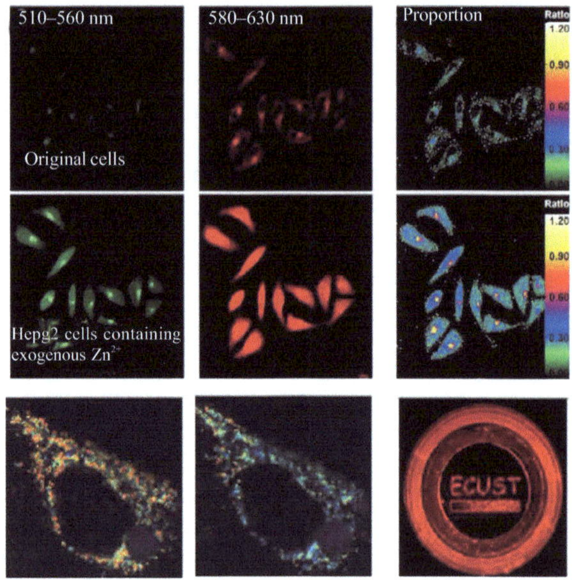

A light-controlled gene expression system for precise intervention of diseases

Fig. 3.2 Imaging molecular probe

microscopy to further improve live-cell imaging [11]. They **solved the crystal structure of tetrameric mEos2** that had yielded the best localization precision achieved thus far for a PAFP, and according to the analysis results, **rationally designed improved versions, mEos3.1 and mEos3.2. As compared with other EosFP variants, mEos3.2 gave the best overall performance in brightness, maturationrate, pH stability, photon budget after switching, and label density. Therefore, it could be considered as a new probe to replace other EosFPs in both diffraction-limited and super-resolution microscopy** (Fig. 3.2). The development of this ultra-bright photoactivatable fluorescent protein is of great significance to study the precise spatiotemporal regulations of signal transduction, understand the occurrence and development mechanisms of related human diseases, and develop efficient means for diagnosis and treatment.

3.2 Investigations on the Roles of Nucleic Acids in Signal Transduction Processes Utilizing Small Chemical Probes

Nucleic acids store and convey genetic information. They are important material basis for the continuation of life. As the genetic material, their functions not only rely on base sequences (i.e. the primary structure), but also are influenced by their special secondary structures and various chemical modifications. The transformation between these different structures or modifications affects and determines downstream life processes, which is the so-called nucleic acid related signal transduction. The flexibility of the nucleic acid backbone provides a broad space for chemical modifications and ligand designs, making it an important target for functional regulations in vivo. Funded by this MRP, research teams have achieved the following important results.

3.2.1 Studies on Signal Transduction Regulation Based on the Ligands of Quadruplex Nucleic Acids

There are interactions between the chiral metallo-supramolecular compound *M*-$[Ni_2L_3]^{4+}$ (Ni-M) and dimeric G-quadruplexes. It was reported that this M-enantiomer (Ni-M) could **bind to dimeric G-quadruplexes 200-fold better than to monomeric G-quadruplex**. The research group of QU Xiaogang from CAS Changchun Institute of Applied Chemistry found that the dimeric G-quadruplexes could serve as a new target for G-quadruplex-interactive ligands, which provided a new way for designing the ligands targeting telomeric G-quadruplex [12]. Given conformational transitions of G-quadruplexes under different pH conditions, the study demonstrated that human telomeric G-quadruplex could switch from mixed conformation to antiparallel structure under acidic conditions. Such transitions were resulted from the conformational changes of G-quadruplex loops, and the G-quadruplex with central TT loop was prone to undergo such changes under acidic conditions [13]. The group also systematically expounded the biological activity of SWCNTs, the ligands that could selectively stabilize i-motif. For the first time, they proved that SWCNTs could inhibit telomerase activity through induced formation of i-motif and G-quadruplex, to interfere with the structures and functions of telomeres, resulting in cell cycle arrest, apoptosis or senescence. This might be caused by telomere structure dysfunction, which were closely related to the upregulation of p16 and p21 proteins. Their work **has provided new insights into understanding the biomedical effects of SWCNTs in cancer cells and the biological importance of i-motif structure in vivo** [14].

The research group of SONG yangzhou from Sun Yat-Sen University reported that at late S/G2/M phase of human cells, TPP1 could recruit more telomerase to telomeres, which was regulated by the phosphorylation of S111 within the TPP1. **The structural analysis indicated that the phosphorylated Ser111 (S111) resided in the TPP1 OB domain, interacting with telomerase**. Mutating Ser111 to Ala or other amino acids affected the recruitment of TPP1 by telomerase, which led to telomere shortening. Their findings provide the research basis of signal transduction and structural biology for understanding the mechanism of **controlling cell cycle-dependent telomerase recruitment and telomere elongation through phosphorylation of TPP1** [15]. The 4,4'-dihydroxyazobenzene compounds synthesized by the research group of ZHOU Xiang from Wuhan University could regulate the conformation of human telomere DNA by light [16]. A further study indicated that this azobenzene and its derivatives **interconverted between the *trans* and *cis* forms under alternate 365 nm UV and visible light (VIS) irradiation.** Changes in the structure or conformation of enzymes were often accompanied by changes in their catalytic activity. These conformational changes might **result from the binding/dissociation of different regulators**. The azobenzene derivative **interconverted between the *trans* and *cis* forms under alternate 365 nm UV and VIS (>400 nm) irradiation, which consequently triggered the compaction or extension of the DNA aptamers**, to affect the binding of the aptamer and enzyme, regulating the enzyme activity. They selected the thrombin as the enzyme to be regulated, studied its catalytic activity in vitro, and successfully used this stragety to **effectively control blood clotting in human plasma**. Compared with previous strategies, their approach is easy and cheap, and it is **potentially beneficial in future biomedical and pharmaceutical applications** [17].

G-quadruplex (G4) is one of the most important secondary structures in nucleic acids. However, G4 RNAs had not been reported in any RNA viruses, such as the hepatitis C virus. The bioinformatics analysis of ZHOU Xiang's group **revealed highly conserved guanine-rich consensus sequences within the *core* gene of hepatitis C despite the high genetic variability** of this RNA virus. They **further showed using various methods that such consensus sequences could fold into unimolecular G4 RNA structures, both in vitro and under physiological conditions**. They also **provided direct evidences that small molecules specifically targeting G4 could stabilize this structure to reduce RNA replication and inhibit protein translation of intracellular hepatitis C. The stabilization of G4 RNA in the genome of hepatitis C represents a promising new strategy for anti-hepatitis C drug development** [18].

3.2.2 Regulation Based on Modified Nucleic Acids and Non-canonical Nucleic Acid Structures

Given the diversity of nucleic acid structures, it is important to recognize and regulate nucleic acids of different structures with small molecules to reveal many life processes. Nucleic acid changes are also important causes for many major diseases. The small-molecule recognition and regulation of nucleic acids provide important theoretical basis and technical methods for the diagnosis and treatment (drugs) of these diseases. ZHOU Xiang's group synthesized the ruthenium complex $[Ru(dip)_2dppz]^{2+}$, which could induce the right-handed to left-handed transition of various DNA sequences at low salt concentrations, such as full-CG, full-AT, non-alternating pyrimidine-purine, and genomic sequences. This complex induced the Z-DNA conformation, breaking the limit of original sequences, and the left-handed helix DNA structures of various sequences (that were stable and bioactive) were successfully observed under approximately physiological conditions [19].

ZHOU Xiang's group found that different cytosine modifications (including methylation, hydroxymethylation, formylation, and carboxylation) on CpG island sequences had different effects on the B-Z transition of DNA. They reported that 5-methylcytosine (5mC), 5-formylcytosine, and 5-carboxylcytosine facilitated the transition, while 5-hydroxymethylcytosine (5hmC) inhibited it, which is of epigenetic values [20]. The development of detection techniques of modified nucleic acids is an important area of epigenetic research, and identifying them with small chemical molecules is a crucial frontier. ZHOU Xiang's group synthesized N-halogeno-N-sodiobenzenesulfonamide reagents to recognize 5mC in DNA sequences in a highly selective manner, and distinguish 5mC from 5hmC and 5-formylcytosine. This is expected to be incorporated into the next generation sequencing method to provide a powerful tool for research in epigenetics and cancer diagnosis [21]. They also developed the technique of detecting 5-formylcytosine in the genome [22].

The modifications of nucleic acids could directly affect related biological functions. According to the study of the research group of ZHOU Songyang, in cells with hypomethylated genomes, the Daxx/Atrx protein complex **preferentially translocated to tandem repetitive elements** and recruited Suv39h to promote H3K9 trimethylation, to ensure the specific silencing of the repetitive elements in the genome, including telomeres, to prevent gene mutation induced by random insertion of transposons. Their results reveal a new mechanism of epigenetic regulation of special histone modifications [23]. The process of Vitamin C participating in multiple epigenetic regulations to promote reprogramming is shown in Fig. 3.3.

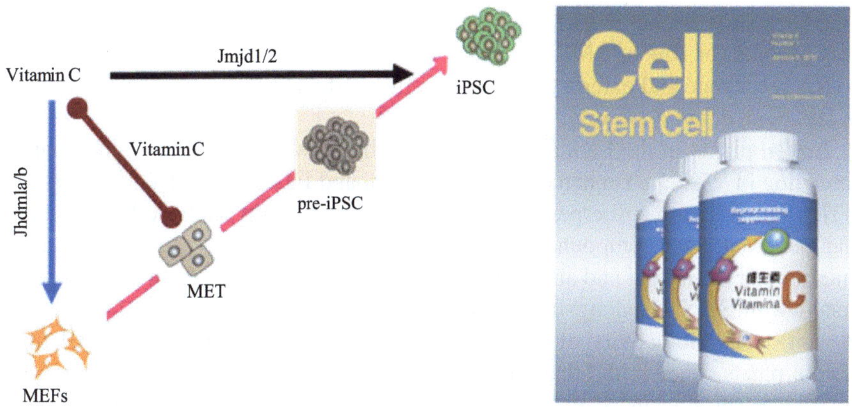

Fig. 3.3 Vitamin C participates in multiple epigenetic regulations to promote reprogramming

3.3 Revealing New Mechanisms of Cell Signal Transduction Processes with Active Natural Products as Probes

It has become a focus of chemical biology research to utilize active small-molecule compounds as probes for studying gene/protein regulations and functions, to reveal molecular mechanisms and signal transduction networks of important cellular life processes, and to discover new active compounds for regulating cell activities. This has also improved the abilities for studying complex biological pathways and processes. Targeting on acute myeloid leukemia (AML), the research group of CHEN Guoqiang from Shanghai Jiao Tong University focuses on the core scientific theme of "exploring and expounding the molecular events in signal transduction pathways with small-molecule compounds", and has carried out a series of studies utilizing the strategies such as "discovering anti-leukemic active compounds" "exploring related drug targets" and "revealing molecular mechanisms of leukemia cell fate determination".

3.3.1 Discovery of Anti-leukemic Active Compounds

AML is a group of heterogeneous hematological malignancies, and it contains various subtypes. During the past 20 years, great progress has been made in the research of AML-related pathogenesis and treatment. One important advance is to treat the acute promyelocytic leukemia (APL) with all-trans-retinoic acid (ATRA) and arsenic trioxide (As_2O_3). ATRA can induce remission in most APL patients to enable their long-term survival, but it has the problem of drug resistance. However, APL patients only account for 10% of AML patients. It is thus important to study how to overcome ATRA resistance and extend the differentiation-inducing therapy to other subtypes of leukemia or solid tumors.

Isodon plants in the Lamiaceae family are herbs widely used in China, mainly spreading in Western Yunnan, Western Sichuan, and Eastwestern Tibet. This genus is rich in ent-kaurene diterpenoids, from which over 700 types have been isolated and identified. Such compounds usually have biological activities, for example, anti-inflammatory, antiviral and antibacterial properties. CHEN Guoqiang's group cooperated with SUN Handong from CAS Kunming Institute of Botany on studying the anti-leukemic activity of natural small-molecule compounds originated from *Isodon* plants. They found that pharicin B could stabilize the retinoic acid receptor α (RAR-α) protein in AML cell lines and primary cells from AML patients, as well as the promyelocytic leukemia (PML)-RAR-α protein, but had no effects on the ATRA-induced cleavage of PML-RAR-α protein. Through such stabilization, pharicin B could enhance the ATRA-dependent transcriptional activity, including C/EBP β and RIG-G. Based on that, they discovered that pharicin B could significantly increase ATRA-induce cell differentiation in AML cell lines U937 and NB4, and primary AML cells isolated from newly diagnosed patients, and it could restore ATRA sensitivity in some retinoic-acid-resistant APL cells NB4-LR1 [24].

The study on adenanthin, the diterpenoid isolated from *Isodon adenanthus*, is particularly interesting. They found that the adenanthin of high concentrations were cytotoxic, but under low concentrations, it could induce APL cell line NB4 cells and primary leukemia cells from AML patients of multiple subtypes to experience morphological and functional differentiation. Such differentiation-inducing effect occurred not only in the ATRA-sensitive NB4 cells, but also in ATRA-resistant cells, including NB4-LR1, NB4-LR2, and MR2 cells. They then conducted in vivo study on the ATRA-sensitive and ATRA-resistant AML mice respectively derived from APL-specific PML-RAR-α and mutated PML-RAR-α transgenic mice. It was found that for both mice, adenanthin could effectively **induce the differentiation of APL-like cell** in vivo, repressed tissue infiltration in the mice, and thus significantly prolonged the survival of the APL mice. These important findings indicated that the differentiation-inducing effect of adenanthin was different from that of ATRA. In addition, they found that adenanthin did not destroy PML-RAR-α fusion protein or regulate its distribution in cell.

3.3.2 Discovery of Adenanthin Targeting Peroxiredoxin I/II

CHEN Guoqiang's group assumed that the α,β-unsaturated ketone of adenanthin was its active moiety. It formed a covalent complex with its targeting proteins through the Michael addition reaction. When the double bond of adenanthin was reduced to a single bond, its differentiation-inducing effect would be completely abolished. However, the modification of the C3 hydroxyl group did not affect its differentiation-inducing activity. They then **installed a biotin tag to its C3 hydroxyl group to yield biotin-adenanthin** through chemical conjugation. This compound could still induce the differentiation of AML cells efficiently. Next, **NB4 cell lysates were incubated with** biotin-adenanthin, and the proteins that biotin-adenanthin might bind were screened through immunoprecipitation. According to the results, one protein band was precipitated by biotin-adenanthin, and it was competed away by adenanthin, **indicating that the protein that bounds biotin-adenanthin also bounds adenanthin**. The mass spectrum **analysis revealed that this band was peroxiredoxin (Prx) I or Prx II**.

Prxs are a family of peroxidases of catalase activities. There are at least 6 Prx members in human, namely Prx I–VI. Prxs **have a conserved cysteine** (Cys) named peroxidatic cys (CP) that **serves as the site of oxidation by peroxides**.

Peroxides oxidize the Cys-SH of CP to Cys-SOH, **which reacts with another Cys typically named the resolving cysteine** (CR), forming a disulfide to remove the peroxides. **Emerging data have implicated an association** between the expressions of Prxs and **major human diseases such as cancers, cardiovascular diseases, diabetes and neurodegeneration**. Further studies indicated that adenanthin did not bind to Prx IV, V, or VI proteins. It had a low affinity binding with Prx III, which was a mitochondrial protein, while adenanthin did not exist in the organelle. Therefore, they focused on exploring the implications of adenanthin binding Prx I and Prx II. Interestingly, repressing Prx I or Prx II expression with siRNA directly induced the differentiation of leukemia cells. The following protein mass spectrometry analysis and site mutation study indicated the specifically covalent binding of adenanthin to the CR sites of Prx I and Prx II. Accordingly, it obviously inhibited the enzymatic activity of Prx I and Prx II, especially Prx I.

3.3.3 Signaling Pathways of Adenanthin Targeting Prx I/II to Induce AML Cell Differentiation

The inhibition of enzymatic activity of Prx I/II by adenanthin increased intracellular H_2O_2. In the presence of N-acetyl-L-cysteine (NAC), a H_2O_2 remover, the differentiation-inducing effect of adenanthin was almost inhibited, indicating the important role of H_2O_2 in this process. The damage effect of H_2O_2 on cells is well understood that it could cause damages to cells with the oxidation of lipid, protein, DNA and other biomacromolecules. However, mounting evidence has indicated that acting as a second messager, it is widely involved regulating cellular signaling pathways and various life processes. The understanding of its important role in the processes of normal and abnormal hematopoiesis is growing. Further studies have indicated that increased H_2O_2 activated extracellular signal-regulated protein kinases (ERK1/2), upregulating the expression of hematopoietic cell differentiation related transcription factor C/EBPβ, to induce cell differentiation.

3.3.4 Discovery of New Compounds Inducing AML Cell Differentiation with Prx I as a Target

Based on the research mentioned above, CHEN Guoqiang's group utilized Prx I as the target, and through computer virtual screening, in vitro activity assay, and intracellular interaction analysis, they found that the small molecule H7 could inhibit Prx I in a noncovalent manner. H7 treatment increased reactive oxygen species (ROS) level in cell and induced NB4 cell differentiation. Therefore, H7 induced cell differentiation through the ROS-ERK1/2-C/EBP pathway, and the leukemia-cell differentiation in mice [25].

3.3.5 *Extended Research on the Antitumor Effect of Aden-Anthin Targeting Prx I/II*

Different members of the peroxiredoxin family are expressed abnormally in various types of tumors, so they are regarded as potential targets for the drug screening of cancer chemotherapy. CHEN Guoqiang's group found that at certain concentrations (3–9 μmol/L), adenanthin could induce the apoptosis of hepatocellular carcinoma (HCC) cells in a time- and dose-dependent manner, without affecting the activity of normal liver cells. Complete silencing of Prx I led to spontaneous apoptosis of HCC cells without affecting normal liver cells; partially silenced Prx I and Prx II increased the sensitivity of HCC cells to the apoptosis induced by adenanthin, but had no effect on the resistance of normal cells to the apoptosis induced by a certain dose of adenanthin. The restored expressions of siRNA-resistant Prx I/II in the partially silenced Prx I/II HCC cell lines led to restored sensitivity to the apoptosis induced by adenanthin. In animal experiments, both 10 and 20 mg/kg adenanthin significantly suppressed the tumor growth in the mice with tumor burden. Their findings indicate that the ROS accumulation led by Prx I/II suppression can induce not only the leukemia cell differentiation, but also the apoptosis of HCC cells selectively, without affecting the activity of normal liver cells. This has provided lab evidence for the development of Prx I- and Prx II-targeted antitumor drugs [26].

Their study (Fig. 3.4) has reported the first active lead compound targeting Prx protein, and also the role of inhibiting Prx I/II protein activity in the leukemia cell differentiation. The related work was published in *Nature Chemical Biology*, and recommended as a featured article on the cover [27]. It has been summarized as one of the 10 successful cases of drug target identification by Prof. Waldman et al. of Max Planck Institute for the Advancement of Science in German, and selected as one of the Top 10 Scientific Advances in China in 2012.

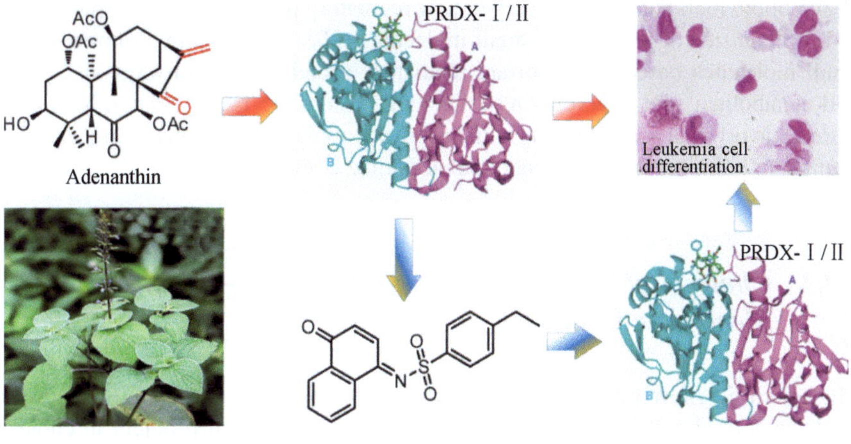

Fig. 3.4 Discovery of adenanthin targets and study on the molecular mechanisms

3.3.6 Regulation Mechanism of Wnt Signaling and Apoptosis Pathways by the Natural Product Diterpenoid S-3

The research group of LI Lin from CAS Institute of Biochemistry and Cell Biology utilized the previously discovered natural product NC043/S3 as a probe to identify its target protein to specifically modulate the Wnt signaling pathway, and defined its structure–activity relation of inhibiting the pathway. They further optimized this small molecule to produce a simple and effective Wnt signal inhibitor [28]. Additionally, the group found that NPHXL, the target protein of NC043/S3, could promote the transcriptional activity of the classic Wnt signaling pathway, and act on the β-catenin downstream to enhance the binding between β-catenin and TCF. The research group of CHEN Quan from CAS Institute of Zoology identified the compound S-3 via the screening with Bax/Bak double knock-out murine embryonic fibroblasts (MEFs), and found that it induced Bax/Bak-independent apoptosis; they also reported the molecular mechanism that induced apoptosis by altering the conformation of Bcl-2 [29].

3.4 Investigations on Regulation Mechanisms of Glycolipid Metabolism via Small-Molecule Probes

Cardiovascular diseases, diabetes, obesity, and fatty liver have become major diseases threatening both physical and living quality of the population in China. The glycolipid metabolism disorder is closely related to these diseases. One prominent feature of glycolipid metabolism is the interactions and regulations among various metabolic small molecules and biomacromolecules, such as proteins. Therefore, the studies utilizing small molecules have unique advantages. The research group of SONG Baoliang from Wuhan University focuses on important pathways in the regulation of glycolipid metabolism: discovery of new regulatory proteins, signaling pathways, and mechanisms of action with small-molecule probes, identification of new active small molecules based on important regulatory proteins and pathways of glycolipid metabolism, and intervention of metabolic disorders. Their research has made systematic original achievements with promising application prospects, and exerts a significant influence in the international academic community.

3.4.1 Cholesterol Transport Through Lysosome-Peroxisome Membrane Contacts

Cholesterol is unevenly distributed and dynamically transported in cells, which is vital for normal cellular life processes. This is a fundamental problem in cholesterol metabolism, and also a basic problem in cell biology. SONG Baoliang's group utilized the human genome-wide lentiviral shRNA library, and designed an efficient screening strategy of specific enrichment of intracellular cholesterol trafficking defective cells to identify over 300 genes affecting cholesterol transport. **They then found that genes encoding peroxisomal proteins were statistically enriched**. The silencing of ABCD1, BAAT, ACOT8, PEX1, PEX3, PEX6, or PEX10 led to defects in the intracellular cholesterol transport. Moreover, the cholesterol accumulation **was observed in lysosome, but not peroxisome**. They selected ABCD1 for further studies. Immunofluorescence assay indicated that there was about 20% contact between ABCD1 and lysosome-peroxisome. They named this newly-discovered phenomenon as lysosome-peroxisome membrane contact (LPMC). To further verify the lysosome-peroxisome relationship, they used different strategies such as super resolution structured illumination microscopy (SR-SIM), PerkinElmer UltraVIEW VoX 3D system imaging, transmission electron microscopy imaging, time-lapse live-cell imaging, organelle immunocoprecipitation, and in vitro reconstitution. All the experimental results confirmed the presence of LPMC. They also found that interference with NPC1 or ABCD1 expression significantly decreased the LPMC. It depended on the level of intracellular cholesterol, and the low density lipoprotein (LDL) originating from cholesterol could enhance this dynamic contact.

To further reveal the lysosome-peroxisome contact mechanism, **multiple proteomics experiments were employed to analyze lysosomal membrane proteins, peroxisomal proteins, and NPC1 interacting proteins. After merging of the protein lists, the proteins involved in vesicle fusion or organelle dynamics were selected as refined candidates**, among which synaptotagmin VII (Syt7) drew their attention. As a type of lysosomal membrane protein, its knockdown caused cholesterol accumulation in lysosome, and **the LPMC was also dramatically diminished**. The in-vitro binding assay showed that Syt7 was crucial to the LPMC. These findings indicated that Syt7 was **a lysosomal protein required for LPMC formation**.

They finally summarized **the model for cholesterol transport from lysosome to peroxisome. After internalization, LDL particles were delivered to lysosome** where they were hydrolyzed to free cholesterol by acid lipase. In the lysosomal cavity, **the NPC2 protein bound free cholesterol with the 8-carbon isooctyl side chain**, and the exposed OH group and molecular scaffold were then identified and bound by the N-terminal domain of NPC1. **The N-terminal domain of NPC1 could penetrate the glycocalyx and facilitate cholesterol to insert into the lysosomal membrane. Lysosome and peroxisome then formed close membrane contacts through interaction between Syt7 and P1(4, 5)P2. Thus, cholesterol could move from lysosome to peroxisome**. This work was published in *Cell* in April 2015 [30], and selected as one of China's top 10 research advances in life sciences in 2015.

3.4.2 Negative Regulation Mechanism of Cholesterol Synthesis Metabolism and Molecular Pathway of Intestinal Absorption of Cholesterol

As an important negative regulation pathway, the degradation of HMGCR protein mediated by sterol can control the endogenous synthesis of cholesterol. SONG Baoliang's group systematically studied the degradation of HMGCR through the ubiquitin–proteasome pathway. They found that sterol specifically promoted the ubiquitination of HMGCR, and identified the ubiquitin ligase gp78 and its complex. Their further study with the liver-specific gp78 knockout mice indicated that gp78 not only mediated HMGCR degradation, but also promoted the degradation of Insig, a negative regulator of the SREBP pathway. These mice had increased HMGCR and Insig, by which the SREBP pathway was suppressed, finally leading to decreased cholesterol synthesis. They also **produced more FGF21, which activated thermogenesis in brown adipocytes, enhanced energy expenditure**, and improved obesity and insulin sensitivity. The group found that betulin could inhibit the endogenous synthesis of lipids, reduce the levels of cholesterol and triglyceride, improve body metabolism and prevent metabolic diseases [32].

3.4.3 Exploration of Endoplasmic Reticulum Stress and Glycolipid Metabolism and Discovery of Important Regulatory Active Small Molecules

The endoplasmic reticulum (ER) is a critical location for the synthesis, folding, and quality control of proteins in cells. Its homeostasis is vital for maintaining normal cell functions. ER stress, which results from the failure of ER in coping with the load of misfolded proteins, initiates the signaling pathway of unfolded protein response (UPR). The three classic UPR pathways, mediated by ER transmembrane proteins IRE1α, PERK, and ATF6 respectively, play a pivotal role in regulating glycolipid homeostasis. The research group of LIU Yong from the Institute for Nutritional Sciences of CAS Shanghai Institutes for Biological Sciences found, in the previous studies with animal models, that the protein kinase/ribonuclease IRE1α played an important role in the glycolipid metabolism disorder. Therefore, they screened the compound library with the luciferase reporter assay of IRE1α-dependent splicing of XBP1 mRNA. Among the screened candidate compounds for inhibiting the ribonuclease activity of IRE1α, two new potential inhibitors WNN1287-H011 and DCDS0047-H011 were discovered and validated with cell detection systems. However, both were found to be of strong cytotoxicity later. They thus synthesized 4U8C and MKC-3946, inhibitors of the known IRE1α nuclease, and further performed proof of concept studies in animal models of obesity and diabetes. Additionally, given the regulation of FXR on ER stress, they carried out studies on screening, discovery, and structural transformation of highly active FXR antagonist compounds, as well as their functions and mechanisms of lowering blood glucose and cholesterol.

LIU Yong's group found in their study on pancreatic β cells that the scaffold protein RACK1 **interacted with IRE1α in a glucose-stimulated or ER stress-responsive manner**. RACK1 was constitutively associated with protein phosphatase PP2A to control IRE1α phosphorylation, and acted like a "molecular brake" in the physiological dynamic regulation of IRE1α signaling pathway. The prolonged failure of this regulation **might bring about deleterious effects on β cell functions, contributing to the pathological development of** type 2 diabetes [33]. Later, they found that in the tissue-specific knockout mice, the IRE1α signaling pathway played an important regulatory role in the compensatory proliferation of pancreatic β-cells under the obesity-induced metabolic stress [34]. In addition, their study with animal models indicated that hepatic IRE1α pathway could sense energy changes in body and was activated by phosphorylation upon fasting.

Glucagon-stimulated protein kinase PKA directly phosphorylated IRE1α, which was activated to participate in regulating the hepatic gluconeogenesis by glucagon. Under metabolic stress, **IRE1α was highly hyperactivated in mice with obesity**, and **silencing IRE1α markedly reduced hyperglycemia and improved glucose intolerance**. Their findings demonstrate that IRE1α is a sensor of both protein folding and nutrition metabolic signals, indicating a potential treatment of type 2 diabetes by specific modulation of the hepatic IRE1α activity [35]. Their study in liver-specific IRE1α knockout (LKO) mice indicates that hepatic IRE1α functions as a sensor of catabolic signals, and plays an important regulatory role in **promoting the adaptive shift of energy utilization** (from utilization of glucose to fatty acids) **during starvation** [36].

LIU Yong's group chose the inhibition of ER stress as the primary screening index, obtained, designed, and synthesized 69 antagonists from over 10,000 compounds with the two modes of virtual screening and pharmacological random screening. Based on that, the group explored their structure–activity relations, activities for inhibiting ER stress and related mechanisms of action at the levels of molecules and cells. The candidate compound Cpd15 was tested for its sugar- and lipid-lowering efficacy in animal models, and after structural optimization, its hypoglycemic effect was further examined in mice with obesity. The group thus verified the possibility of utilizing FXR and ER stress signaling pathways to discover and obtain the compounds with effects of lowering both blood glucose and cholesterol.

3.5 Discovery and Validation of Drug Targets and Lead Compounds Based on Signal Transduction Processes

In terms of original innovation, discovery of new drug targets is of more scientific value and greater practical significance than compound innovations, because it is a source of blockbuster innovative drugs, providing opportunities for rapid and efficient discovery of new drugs. Target discovery and functional validation are thus important tasks for research of functional genomics and innovative drugs. As illustrated by practice in recent years, they are challenging, mainly due to the difficulties in target validation. Traditional technologies involved in target discovery and validation are genomic technologies which feature large investments, long cycles, and high risks. Accordingly, they are abandoned by most research institutions and pharmaceutical companies for discovering and validating targets.

Chemical biology can overcome such shortcomings. Technologies of this discipline are much more reliable than genomic technologies for these applications. More importantly, they enable target discovery and validation to go side by side with drug discovery and development. That is, they help to obtain physiological functions of the gene products and the active compounds regulating their activities at the same time. Therefore, this MRP has established the research direction of "discovery of important markers, targets and lead compounds based on signal transduction processes". Targeting at major diseases, it aims to explore functions and interaction networks of important proteins involved in signal transduction processes, and discover important disease-related markers, drug targets, and lead compounds, to lay the foundation for disease prevention, diagnosis and treatment. Researchers engaged in this research direction have given full play to China's advantages in small-molecule compound library synthesis, phytochemistry, and computational chemistry. Based on the research results of the signal transduction of this MRP, they have cooperated with pharmaceutical experts in studies on functional validation of new target druggability and discovery of lead compounds for several major diseases. They have formed distinguishing features in the research through the integration of investigations on signal transduction processes and target discovery, target discovery and functional validation, target functional validation and active compound discovery.

3.5.1 Discovery and Functional Validation of Targets and Design of Lead Compounds for Antitumor Drugs

(1) **New targets of antitumor drugs—copper chaperones Atox1 and CCS**

Disordered copper ion levels lead to various diseases, for example, tumor development. The concentrations of copper ions in the plasma of clinical cancer patients are related to the tumor development and their responses to treatment. Copper chelators (e.g. tetrathiomolybdate) are the only way to regulate the level of copper ions in the body. They can also chelate other metal cations nonspecifically, so they often cause severe toxic side effects. There are various copper chaperones, such as Atox1 and CCS, in charge of cellular copper transport. Their expressions in numerous tumor tissues significantly exceed those of normal tissues. They are also highly expressed in various leukemia cell lines and patient samples. This information indicates that the survival of tumor cells may depend on these copper chaperones. The research group of JIANG Hualiang from CAS Shanghai Institute of Materia Medica cooperated with international teams, including HE Chuan's team of the University of Chicago, on the chemical biology research of the targetability of copper chaperones.

With a computational drug design method, they screened from the compound library the compounds that could bind to the copper-transfer interfaces of Atox1 and CCS. A similarity search was performed based on the feedback from bioactive tests, to find the compounds with stronger capacities of binding and drug-like characteristics. The subsequent biochemical and cell biological experiments indicated that some compounds blocked the cellular copper transport through binding to the copper-transfer interfaces of Atox1 and CCS, to specifically inhibit tumor cell proliferation without affecting normal cells. Of these compounds, DCAC50 showed the best antitumor effect with the least toxicity. Further mechanism investigations demonstrated that it interfered with the copper transport to suppress the activity of Cu/Zn superoxide dismutase (SOD1) with copper as a cofactor, increased the ROS levels in tumor cells, and affected mitochondrial functions to reduce ATP production. The tumor cells were under high oxidative stress, so they were more sensitive to DCAC50. Their findings have not only provided excellent lead compounds for tumor targeted therapies, but also opened up a new area for research on antitumor strategies [37].

(2) **New target for renal cancer treatment—ubiquitin ligase SPOP**

Currently, the first-line drugs for the treatment of renal cell carcinoma mainly target tyrosine kinases. These inhibitors are generally low in the objective response rates and prone to drug resistance. Therefore, it is vital to develop anti renal cancer drugs of new targets. It has been reported that in the clear-cell renal cell carcinoma (ccRCC) under hypoxic microenvironment, the accumulation of HIF leads to the overexpression of the speckle-type POZ protein (SPOP), an adaptor of ubiquitin ligase E3. Meanwhile, such microenvironment would cause its dislocation in cells, transferring from the nucleus to the cytoplasm. SPOP in the cytoplasm can promote the degradation of its substrate proteins—combinative phosphatase and tensin homolog (PTEN) and dual specificity phosphatase 7 (DUSP7), enhancing cell proliferation and tumorigenesis. However, there had been no reports of SPOP inhibitors thus far. JIANG Hualiang's and YANG Caiguang's groups from CAS Shanghai Institute of Materia Medica cooperated with LIU Jiang's group from CAS Beijing Institute of Genomics, and discovered small-molecule inhibitors disrupting the interactions between SPOP, PTEN, and DUSP7. With the method combining virtual screening and in vitro biochemical analyses as well as structural transformation and optimization based on the structure–activity relations, they obtained the active compound Cmp6b that disrupted the interactions between SPOP and its substrate proteins PTEN and DUSP7 in vitro. Through the intracellular experiments of immunocoprecipitation (Co-IP) and ubiquitination assay, they found that SPOP inhibitors could interfere with the SPOP-substrate interactions at a low concentration, and inhibit the ubiquitination of PTEN and DUSP7 by SPOP. As a result, the accumulation of both substrate proteins were observed in ccRCC cells, which **subsequently led to increases in phosphorylated Akt and ERK** in the downstream signaling pathways, resulting in the death of ccRCC cells. In the expriments with the nude mice bearing tumor cell

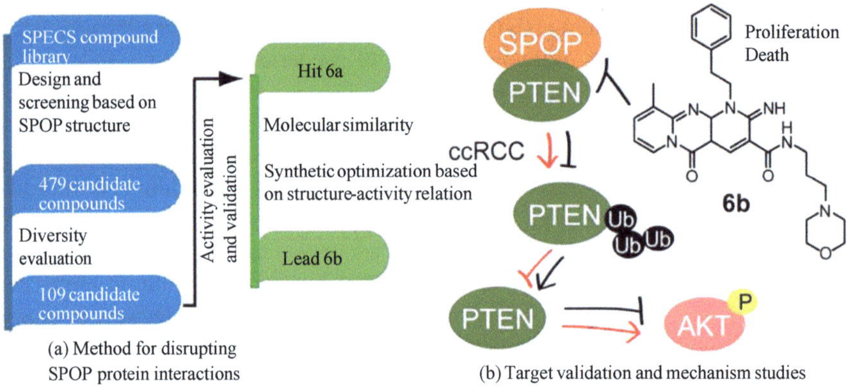

Fig. 3.5 Small molecules target SPOP in renal cancers

xenografts, they found that Cmp6b could significantly kill ccRCC cells and reduce tumor sizes in the mice. The research process of small molecules targeting SPOP in renal cancers is shown in Fig. 3.5. This research has convincingly shown that disrupting SPOP-substrate interactions could be a new strategy for specific targets against renal cancers [38], which will promote the discovery and development of innovative drugs in treating renal cancers.

(3) **New drug-resistant pathway of melanoma and novel antitumor drugs of induced autophagy**

In the previous study [39], the research group of WU Qiao from Xiamen University obtained compound 1-(3,4,5-Trihydroxyphenyl)nonan-1-one (THPN) through screening, which could specifically inhibit the occurrence and development of melanoma through the autophagic cell death mediated by TR3 (Nur77). To expand the application of THPN, they studied the molecular mechanism of the insensitivity to autophagic death induced by THPN and its derivatives in other types of cancer cells (for example, cervical cancer cells and gastric cancer cells). The activity level of Akt in the nonmelanoma cancer cells was clearly higher than that in melanoma cells, resulting in the inhibition of the autophagic cell death signaling pathway induced by THPN and its derivatives. The reason was that the phosphorylation of TR3 at Ser533, by the highly phosphorylated Akt2 in tumor cells, **impeded TR3 nucleocytoplasmic translocation and THPN-induced interaction with Nix**, as well as the subsequent translocation to mitochondria and the autophagic cell death caused by mitochondrial dysfunction. The reduction in the cellular expression level of Akt2 by Akt inhibitors or gene therapies could enable tumor cells to respond to the autophagic cell death induced by THPN and its derivatives. Based on that, WU Qiao's group utilized THPN as a parent structure, explained the THPN-TR3 cocrystal structure for further optimization of compounds, and synthesized a series of THPN derivatives. They finally obtained a compound that was more effective than THPN in inhibiting tumor growth through the induced autophagic pathway. Their work has identified the molecule interfering with THPN-induced autophagic death of tumor cells, expounded a new pathway for TR3 and Akt to regulating their drug resistance, and also provided important theoretical basis for broadening the application of THPN and developing novel drugs for cancer therapy by the induction of autophagic cell death [40].

3.5.2 Discovery and Validation of a New Target for Anti-Inflammatory Drugs

Appropriate inflammatory reactions can remove pathogens while excessive ones may lead to organ dysfunction, failure, and even death. Therefore, illumination of precise regulatory mechanisms and signaling pathways is key to treating various inflammation-related diseases. Sepsis is a systemic inflammatory response syndrome resulted from an infection. **Severe sepsis may result in organ failure and even death**.

The research group of XU Qiang from Nanjing University found that andrographolide (Andro) hampered the assembly and activation of NLRP3 inflammasome and IL-1β release by induced mitochondria autophagy of macrophages, to attenuate colitis progression and finally inhibit the development of colorectal cancer [41]. Further studies indicated that Andro bound to protein p110, which directly interacted with NLRP3 inflammasome to negatively regulate its second signal. It is a new regulation mechanism of inflammasome signaling, providing new insights in the treatment of related inflammatory diseases.

Many anti-inflammatory drugs are developed with the targets of p38 and the associated upstream and downstream proteins of its pathways. They usually have adverse side effects or poor efficacy. WU Qiao's group found that compared with wild-type (WT) mice, Nur77 knockout (KO) mice were more susceptible to the sepsis induced by lipopolysaccharide (LPS) or cecal ligation and puncture (CLP). The expression levels of cytokines were significantly higher in KO mice than in WT mice. All infected mice died within 3 days. These results indicated that Nur77 functioned as an inhibitor in inflammatory reactions. According to their mechanism analyses, effective block of p38α phosphorylation could ensure the unique role of Nur77 in inhibiting inflammation. Based on this mechanism of action, they screened a small-molecule compound PDNPA that specifically targeted Nur77. It bound to the LBD of Nur77 to inhibit p38α from binding to it, **abolishing the phosphorylation of Nur77 by p38α, so that Nur77 can** efficiently inhibit LPS-induced inflammation, thus to effectively treat sepsis and increase the survival rate of mice. Its mechanism of action was completely different from that of the main-stream p38α inhibitors. It did not directly affect the p38α kinase activity. Instead, it targeted Nur77 to impede p38α-Nur77 binding, to keep Nur77 poorly phosphorylated, thus to inhibit inflammation more efficiently. The block of p38α kinase in this way might lower toxic side effects of drugs. The process of PDNPA specifically targeting Nur77 to inhibit inflammation is shown in Fig. 3.6. Therefore, their findings not only reveal the role of Nur77 in

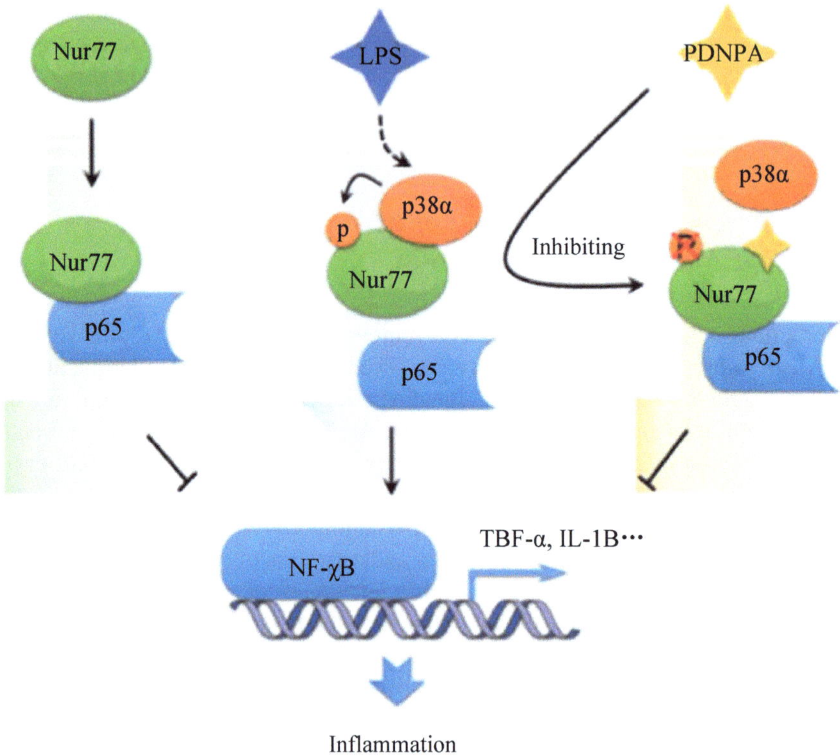

Fig. 3.6 PDNPA specifically targets Nur77 to inhibit inflammation

the negative regulation of innate immune responses and its mechanism of action, but also provide a new target and strategy for developing anti-inflammatory drugs [42].

3.5.3 Studies of Targets in the RNA Epigenetic Signal Transduction Pathways

Epigenetic signal transduction pathways, as an international emerging area of drug target and drug discovery, are characterized by unclear functions of most epigenetic enzyme targets and the lack of small molecules for functional research. Therefore, to discover the first-generation inhibitors or agonists of these enzymes has become an issue of great urgency. YANG Caiguang's, JIANG Hualiang's, and ZHOU Xiang's groups from CAS Shanghai Institute of Materia Medica, as well as JIA Guifang's group from Peking University have focused on the recent frontier in epigenetics—RNA epigenetic enzyme targets, and performed the research and development of the first-generation inhibitors. They have further studied the dynamic modifications for regulating N^6-methyladenosine (m^6A) in mRNA with chemical probes, and provided innovative tools, different from their traditional genetic counterparts, for revealing the signal transduction pathways related to m^6A modifications and cell plasticity.

FTO was the fi RNA demethylase discovered that regulated the modifi abundance of m^6A. However, it was not reported whether small-molecule compounds could target FTO to interfere with m^6A modification in cells. YANG Caiguang's group cooperated with JIANG Hualiang's group, and performed high-throughput virtual screenings based on the structure feature of FTO recognizing m^6A modified bases. They found that the natural product rhein could inhibit the demethylation function of FTO, and interfered with the cellular modification abundance of m^6A in mRNA in a concentration-dependent manner [43]. ALKBH5 was the second m^6A demethylase, and the development of its highly selective chemical probes was a challenge. YANG Caiguang's group worked together with JIA Guifang's group, and designed a high-throughput screening strategy based on the fact that FTO and ALKBH5 had different mechanisms for recognizing methylated nucleic acids. They identified meclofenamic acid (MA), which was a non-steroidal, anti-inflammatory drug, as a selective small-molecule inhibitor of FTO [44]. The influence of MA on the cellular abundance of m^6A was closely related to the expression and functions of FTO, but not to ALKBH5. Based on the structure, they further discovered the molecular mechanism of MA selective inhibition of FTO. Based on their findings, the two groups collaborated with ZHOU Xiang's group, and designed bifunctional chemical probes for visualization of FTO protein [45]. Inspired by the similar chemical structures of MA and fluoresceins, they designed and synthesized the fluorescein derivatives **with the ability to simultaneously selectively inhibit and label FTO**. With techniques of structural biology and click chemistry, they studied the mechanisms of action in detail and validated the targets of the small molecules. Their findings have provided small-molecule tools for "RNA epigenetics" and its related enzymes, and active lead compounds for the discovery and research of new target drugs of related diseases.

3.5.4 Discovery of New Sites and Functions of the Drug Targets for Metabolic Diseases

Purinergic receptors play a vital role in the diseases caused by platelet activation, such as the coronary heart disease and the stroke, which makes the drugs targeting purinergic receptors primary means for the antithrombotic treatment of the coronary heart disease. However, there were few studies on the functions of this receptor family, and their related mechanism remained unclear, due to the low sequence conservation in the family and great difficulties in studying G-protein-coupled receptors (GPCRs). To settle the matter, the research group of ZHAO Qiang from CAS Shanghai Institute of Materia Medica analyzed the 3D structure of P2Y12R bound to AZD1283, and found that the ligand binding pocket of P2Y12R were largely different from those of other class A GPCRs. It was first reported that there were two binding sites in the seven transmembrane helices, indicating that P2Y12R might be able to bind two types of drug molecules simultaneously. Based on its 3D structure information, there were labs crosslinking drug molecules of different types to improve the drug-receptor affinity and reduce side effects.

With the molecular dynamics simulation, the group established the models of interactions between P2Y12R and all marketed drugs and some clinical drugs. The mechanisms of action of most popular antithrombotic drugs targeting P2Y12R, including Plavix, were revealed for the first time. They found with these models that irreversible and reversible P2Y12R drugs had different mechanisms of action, providing evidence for the structure optimization of the two different types of drugs as well as virtual screening and design of new drugs targeting P2Y12R. Additionally, their structural analyses demonstrated for the first time that reversible drugs changed the spatial conformation of the transmembrane helix VI of P2Y12R extracellular domain to inhibit its activity. Their findings could greatly facilitate the future optimization of drugs to improve anticoagulant effects and reduce side effects [46]. The discovery of this new drug binding site is likely to change researchers' understanding of receptor antagonism mechanisms, leading to research on new GPCR functions and new directions of drug development.

The research group of YU Biao from CAS Shanghai Institute of Organic Chemistry cooperated with XIE Xin' s group from CAS Shanghai Institute of Materia Medica, and disclosed that Gordonoside F, a steroid glycoside isolated from *Hoodia*, was a selective agonist of G protein-coupled receptor GPR119. They also explained the mechanism underlying the appetite-suppressing effect of *Hoodia* with in-vitro and in-vivo experiments. It is an original breakthrough in the development of weight-loss drugs with *Hoodia*. The findings on its appetite-suppressing mechanism could provide lead compounds for developing new weight-loss and anti-diabetic drugs, and explain the traditional effects of African plant *Hoodia* [47].

3.5.5 Validation of New Targets and Concepts for Preventing Bacterial Infection

While drug resistance of traditional antibacterials, such as antibiotics, is increasingly severe, development and application of the antibacterials with new mechanisms to avoid resistance are becoming a research hotspot in the prevention of bacterial infection. Supported by this MRP, LAN Lefu's, YANG Caiguang's, and JIANG Hualiang's groups of CAS Shanghai Institute of Materia Medica, as well as LI Jian's group of East China University of Science and Technology have expanded the research ideas of "small molecules regulating cellular signal transduction processes", endeavored to discover new antibacterials with the non-antibiotic mechanism of "disarming pathogens rather than killing them", and performed research on anti-virulence drugs. Unlike traditional antibiotics, anti-virulence drugs aim to achieve the therapeutic effects through blocking pathogenic processes of bacteria rather than directly killing them or inhibiting their growth. The representative achievements are as follows.

(1) **Discovery of the signal transduction based on "sensor" protein CcpE to be potential new anti-*S. aureus* targets**

Staphylococcus aureus (S. aureus) uses the catabolite control protein E (CcpE) to sense its intracellular level of citrate to regulate its metabolic status and expression of various virulence-associated factors, to effectively govern its virulence. LAN Lefu's and YANG Caiguang's groups found that the small molecule citrate was not only an important metabolic intermediate, but also a signaling molecule governing the virulence of *S. aureus*, indicating an intimate link between its metabolic status and bacterial virulence. Additionally, their findings also demonstrated that CcpE was a major negative regulator of controlling its virulence, implying that activation of its function might reduce the bacterial virulence [48].

(2) **Discovery of the new target CrtN for inhibiting *S. aureus* virulence**

LAN Lefu's and LI Jian's groups found, through the phenotypic screening of the marketed drug library, that an anti-fungal drug naftifine significantly **inhibited the production of golden carotenoid pigment in *S. aureus*** at nanomolar concentrations, and **made *S. aureus* more susceptible to** oxidative stress and human blood. Further mechanism studies indicated that naftifine blocked the synthesis of golden carotenoid pigment via the competitive inhibition of *S. aureus* diapophytoene desaturase (CrtN), and **attenuated the virulence of *S. aureus* Newman** strain in mouse infection models. They then demonstrated that CrtN was important in promoting **the virulence of a variety of clinical *S. aureus* isolates, including methicillin-resistant *S. aureus* (MRSA) strains**; accordingly, the CrtN inhibitor (naftifine) could effectively inhibit the infection of mice by the "superbacteria" of important clinical isolates of MRSA strains, including Mu50, USA400, and MW2. Their findings provide a new drug target for developing the alternatives to antibiotics, such as anti-virulence drugs [49]. Based on that, a series of compounds targeting the structure of CrtN protein have been designed. They are of high antibacterial activities, and one of them has been technologically transferred to Hubei Bio-pharmaceutical Industrial Technological Institute Inc. in China for further development.

(3) **Confirmation of SrtA as a target for anti-virulence drugs and discovery of its lead compounds**

The biological functions of sortase A (SrtA) are important to the virulence of gram-positive bacteria. Their SrtA specifically recognizes the signal peptide LPXTG, and anchors multiple virulence factors containing this peptide in the envelope of the pathogen to regulate its infectious virulence to the host. Since there were no good small-molecule inhibitors, it was unclear whether SrtA could serve as a target for anti-infective drugs. Due to the lack of small-molecule probes, chemical intervention methods could not be applied in studies on the mechanisms of SrtA recognizing substrate proteins or functional validation of targets. YANG Caiguang's and JIANG Hualiang's groups worked together, and combined the molecular design of drugs and

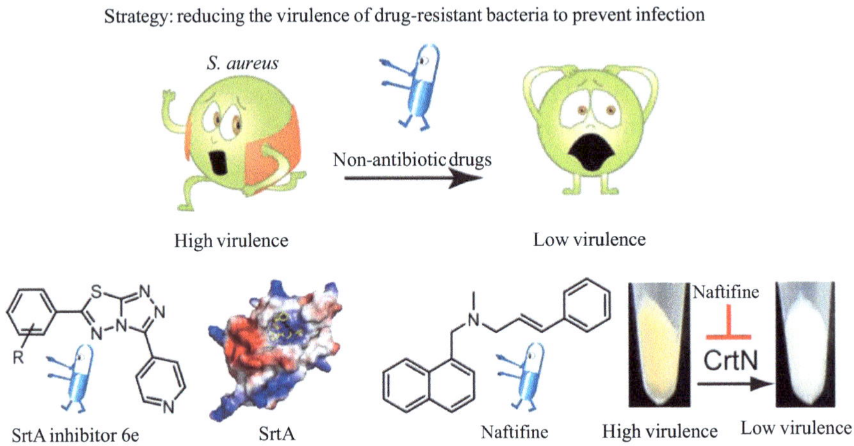

Fig. 3.7 Validation of anti-virulence target

biological screening to identify small-molecule SrtA inhibitors of novel structures. They performed structural optimization and further screens of the active compounds to get inhibitors with higher activities. These compounds could effectively inhibit SrtA from recognizing the surface virulence factors IsdA and SasX, and significantly suppress the transpeptidation reaction of the surface virulence factor SPA catalyzed by SrtA to interfere with its incorporation into the cell wall, resulting in great inhibition of the virulence of drug-resistant *S. aureus*, as shown in Fig. 3.7. These small molecules do not kill bacteria or hardly affect their growth, but they can effectively prolong the survival of the mice infected by drug-resistant bacteria, achieving the goal of "containing infection without killing or inhibiting bacteria". In addition, this skeleton compound can inhibit SrtA of broad- spectrum gram-positive bacteria. Their work has demonstrated the prospect of SrtA as a target for anti-infective drugs at multiple levels of molecules, cells, and animal models [50].

References

1. Li J, Yu J, Zhao J et al (2014) Palladium-triggered deprotection chemistry for protein activation in living cells. Nat Chem 6(4):352–361
2. Wang J, Cheng B, Li J et al (2015) Chemical remodeling of cell-surface sialic acids through a palladium-triggered bioorthogonal elimination reaction. Angew Chem Int Ed Engl 54(18):5364–5368
3. Li J, Jia S, Chen PR (2014) Diels-Alder reaction-triggered bioorthogonal protein decaging in living cells. Nat Chem Biol 10(12):1003–1005
4. Zhang G, Li J, Xie R et al (2016) Bioorthogonal chemical activation of kinases in living systems. ACS Cent Sci 2(5):325–331
5. Li J, Chen PR (2016) Development and application of bond cleavage reactions in bio-orthogonal chemistry. Nat Chem Biol 12(3):129–137

6. Chen X, Tang S, Zheng JS et al (2015) Chemical synthesis of a two-photon-activatable chemokine and photon-guided lymphocyte migration in vivo. Nat Commun 6:7220
7. Lin L, Liu L, Zhao B et al (2015) Carbon nanotube-assisted optical activation of TGF-be-ta signalling by near-infrared light. Nat Nanotechnol 10(5):465–471
8. Wang X, Chen X, Yang Y (2012) Spatiotemporal control of gene expression by a light-switchable transgene system. Nat Methods 9:266
9. Chen X, Li T, Wang X et al (2016) Synthetic dual-input mammalian genetic circuits enable tunable and stringent transcription control by chemical and light. Nucleic Acids Res 44(6):2677–2690
10. Chu TT, Li QQ, Qiu T et al (2014) Clearance of the intracellular high level of the Tau protein directed by an artificial synthetic hydrolase. Mol Biosyst 10(12):3081–3085
11. Zhang M, Chang H, Zhang Y et al (2012) Rational design of true monomeric and bright photoactivatable fluorescent proteins. Nat Methods 9:727
12. Zhao C, Wu L, Ren J et al (2013) Targeting human telomeric higher-order DNA: dimeric G-quadruplex units serve as preferred binding site. J Am Chem Soc 135(50):18786–18789
13. Li W, Wang J, Ren J et al (2013) Near-infrared- and pH-responsive system for reversible cell adhesion using graphene/gold nanorods functionalized with i-motif DNA. Angew Chem Int Ed Engl 52(26):6726–6730
14. Chen Y, Qu K, Zhao C et al (2012) Insights into the biomedical effects of carboxylated single-wall carbon nanotubes on telomerase and telomeres. Nat Commun 3:1074
15. Zhang Y, Chen LY, Han X et al (2013) Phosphorylation of TPP1 regulates cell cycle-dependent telomerase recruitment. Proc Natl Acad Sci USA 110(14):5457
16. Wang X, Huang J, Zhou Y et al (2010) Conformational switching of G-quadruplex DNA by photoregulation. Angew Chem Int Ed Engl 49(31):5305–5309
17. Tian T, Song Y, Wang J et al (2016) Small-molecule-triggered and light-controlled reversible regulation of enzymatic activity. J Am Chem Soc 138(3):955–961
18. Wang SR, Min YQ, Wang JQ et al (2016) A highly conserved G-rich consensus sequence in hepatitis C virus core gene represents a new anti-hepatitis C target. Sci Adv 2(4):e1501535
19. Wu Z, Tian T, Yu J et al (2011) Formation of sequence-independent Z-DNA induced by a ruthenium complex at low salt concentrations. Angew Chem Int Ed Engl 50(50):11962–11967
20. Wang S, Long Y, Wang J et al (2014) Systematic investigations of different cytosine modifications on CpG dinucleotide sequences: the effects on the B-Z transition. J Am Chem Soc 136(1):56–59
21. Wang T, Hong T, Tang T e t al (2013) Application of N-halogeno-N-sodiobenzenesulfonamide reagents to the selective detection of 5-methylcytosine in DNA sequences. J Am Chem Soc 135(4):1240–1243
22. Guo P, Yan S, Hu J et al (2013) Selective detection of 5-formyl-2’-deoxycytidine in DNA using a fluorogenic hydroxylamine reagent. Org Lett 15(13):3266–3269
23. He Q, Kim H, Huang R et al (2015) The Daxx/Atrx complex protects tandem repetitive elements during DNA hypomethylation by promoting H3K9 trimethylation. Cell Stem Cell 17(3):273–286
24. Gu ZM, Wu YL, Zhou MY et al (2010) Pharicin B stabilizes retinoic acid receptor-alpha and presents synergistic differentiation induction with ATRA in myeloid leukemic cells. Blood 116(24):5289–5297
25. Wei W, Ma C, Cao Y et al (2016) Identification of H7 as a novel peroxiredoxin I inhibitor to induce differentiation of leukemia cells. Oncotarget 7(4):3873–3883
26. Hou JK, Huang Y, He W et al (2014) Adenanthin targets peroxiredoxin I/II to kill hepatocellular carcinoma cells. Cell Death Dis 5:e1400
27. Liu CX, Yin QQ, Zhou HC et al (2012) Adenanthin targets peroxiredoxin I and II to induce differentiation of leukemic cells. Nat Chem Biol 8(5):486–493
28. Wang S, Yin J, Chen D et al (2013) Small-molecule modulation of Wnt signaling via modulating the Axin-LRP5/6 interaction. Nat Chem Biol 9(9):579–585
29. Zhao L, He F, Liu H et al (2012) Natural diterpenoid compound elevates expression of Bim protein, which interacts with antiapoptotic protein Bcl-2, converting it to proapoptotic Bax-like molecule. J Biol Chem 287(2):1054–1065

30. CHU B B, LIAO Y C, QI W, et al. Cholesterol transport through lysosome-peroxisome membrane contacts. Cell, 2015, 161 (2): 291–306.
31. Liu TF, Tang JJ, Li PS et al (2012) Ablation of gp78 in liver improves hyperlipidemia and insulin resistance by inhibiting SREBP to decrease lipid biosynthesis. Cell Metab 16(2):213–225
32. Ge L, Wang J, Qi W et al (2008) The cholesterol absorption inhibitor ezetimibe acts by blocking the sterol-induced internalization of NPC1L1. Cell Metab 7(6):508–519
33. Qiu Y, Mao T, Zhang Y et al (2010) A crucial role for RACK1 in the regulation of glucose-stimulated IRE1alpha activation in pancreatic beta cells. Sci Signal 3(106):ra7
34. Xu T, Yang L, Yan C et al (2014) The IRE1alpha-XBP1 pathway regulates metabolic stress-induced compensatory proliferation of pancreatic beta-cells. Cell Res 24(9):1137–1140
35. Mao T, Shao M, Qiu Y et al (2011) PKA phosphorylation couples hepatic inositol-requiring enzyme 1alpha to glucagon signaling in glucose metabolism. Proc Natl Acad Sci USA 108(38):15852–15857
36. Shao M, Shan B, Liu Y et al (2014) Hepatic IRE1α regulates fasting-induced metabolic adaptive programs through the XBP1s−PPARα axis signalling. Nat Commun 5:3528
37. Wang J, Luo C, Shan C et al (2015) Inhibition of human copper trafficking by a small molecule significantly attenuates cancer cell proliferation. Nat Chem 7(12):968–979
38. Guo ZQ, Zheng T, Chen B et al (2016) Small-molecule targeting of E3 ligase adaptor SPOP in kidney cancer. Cancer Cell 30(3):474–484
39. Wang WJ, Wang Y, Chen HZ et al (2014) Orphan nuclear receptor TR3 acts in autophagic cell death via mitochondrial signaling pathway. Nat Chem Biol 10(2):133–140
40. Wang WJ, Wang Y, Hou PP et al (2015) Induction of autophagic death in cancer cells by agonizing TR3 and attenuating Akt2 activity. Chem Biol 22(8):1040–1051
41. Zhang L, Sun Y, Fei M et al (2014) Disruption of chaperone-mediated autophagy-dependent degradation of MEF2A by oxidative stress-induced lysosome destabilization. Autophagy 10(6):1015–1035
42. Li L, Liu Y, Chen HZ et al (2015) Impeding the interaction between Nur77 and p38 re-duces LPS-induced inflammation. Nat Chem Biol 11:339
43. Chen B, Ye F, Yu L et al (2012) Development of cell-active N6-methyladenosine RNA demethylase FTO inhibitor. J Am Chem Soc 134(43):17963–17971
44. Huang Y, Yan J, Li Q et al (2015) Meclofenamic acid selectively inhibits FTO demethylation of m^6A over ALKBH5. Nucleic Acids Res 43(1):373–384
45. Wang T, Hong T, Huang Y et al (2015) Fluorescein derivatives as bifunctional molecules for the simultaneous inhibiting and labeling of FTO protein. J Am Chem Soc 137(43):13736–13739
46. Zhang J, Zhang K, Gao ZG et al (2014) Agonist-bound structure of the human P2Y12 receptor. Nature 509(7498):119–122
47. Zhang S, Ma Y, Li J et al (2014) Molecular matchmaking between the popular weight-loss herb hoodia gordonii and GPR119, a potential drug target for metabolic disorder. Pro Natl Acad Sci USA 111(40):14571
48. Ding Y, Liu X, Chen F et al (2014) Metabolic sensor governing bacterial virulence in staphylococcus aureus. Pro Natl Acad Sci USA 111(46):E4981
49. Chen F, Di H, Wang Y et al (2016) Small-molecule targeting of a diapophytoene desaturase inhibits S. aureus virulence. Nat Chem Biol 12(3):174–179
50. Zhang J, Liu H, Zhu K et al (2014) Antiinfective therapy with a small molecule inhibitor of staphylococcus aureus sortase. Pro Natl Acad Sci USA 111(37):13517

Chapter 4
Outlook

4.1 Domestic Deficiencies and Strategic Needs

The chemical biology discipline in China has come on by leaps and bounds in recent years. Based on this Major Research Plan, great breakthroughs have been achieved in the investigations on signal transduction processes utilizing chemical probes, to realize the strategic goal of "winning a place" in the world. However, the overall competitiveness of China's chemical biology is weaker than that of global scientific powers. Its development still faces many challenges, especially in originality.

The research in life science is going through a period of profound changes. Various fields see breakthroughs in principles and technologies, with hotspots emerging endlessly and changes popping out rapidly. In this context, the research of chemical biology should seize the opportunity, strive for and respond to changes, and solve new biological problems and reveal new biological mechanisms while making a qualitative leap in tool development and technological innovation. Although we have made outstanding achievements in the research of chemical biology, for instance, cellular signal transduction, we have not been sufficiently involved in emerging fields and key directions in biology, and there is no systematic work yet. For instance, chemical modifications of nucleic acids (DNA and RNA) play a vital role in regulating cell fate during the life cycle. However, the categories of chemical modifications of epigenetic regulation, their functions, or relations between dynamic reversible changes and functions have not been well understood. Research on these important chemical modifications mainly relies on traditional biological means. There is a pressing need for the full involvement of chemical biologists to provide strong support for these important biological problems, especially in the development of probe tools for capturing, labeling, and analyzing the chemical modifications and biological functions, as well as analysis techniques for dynamic, real time, and *in situ* detection of changes in these chemical modifications. There is also a need for close cooperation between biologists and chemical biologists to make discoveries and reveal new mechanisms, and even open up new areas of research.

L. Zhang (ed.), *Investigations on Signal Transduction Processes Utilizing Small Chemical Probes*, Reports of China's Basic Research,
https://doi.org/10.1007/978-981-19-9585-9_4

Somatic cell reprogramming is another huge leap in life science. The conversion of somatic cells into pluripotent stem cells can provide them with differentiation potency, which is of great significance. The methods for inducing somatic cell reprogramming with small chemical molecules avoid potential risks of mutation brought by traditional "transgenic" methods as well as the ethical controversy of "nuclear transfer". They are ideal options for controlling cell fate. In particular, Chinese scientists have achieved a breakthrough in the research of chemically induced pluripotent stem cells, which provides a simpler, safer, and more effective method for endowing somatic cells with "pluripotency". It opens up a new way for somatic cell reprogramming. However, research in this field still relies on experiences, with various weaknesses. Especially, the mechanisms of action of small molecules for chemical induction remain unknown, which cannot provide a theoretical basis for further research. It is high time to involve chemical biology in this frontier. We should systematically study the mechanisms of action of these molecules in cells, and reveal their mechanisms by focusing on induced signal transduction regulation and epigenetic changes, to enable the reprogramming of somatic cells with small molecules in a more quantitative, efficient and controllable way. This can contribute to the development and improvement of the theoretical basis for the small molecule induced reprogramming of somatic cells, to provide an ideal cell source for future cell therapy and artificial organs and open the door of regenerative medicine.

We can hardly explain life phenomena without understanding the functions of the brain and the nervous system. Future neuroscience research will focus on molecular mechanisms of signal transduction pathways in key developmental processes and therapeutic methods for major diseases. However, the current involvement of chemical biology in this field is also limited. The rapid development of neuroscience, especially brain science, provides chemical biologists with another great opportunity. Chemical biology will play a vital role in studies on the basic aspects of neuroscience, including functions and complexity of the brain, the molecular basis of cognition and learning, and pathogenesis of neuropsychiatric diseases. Chemical probes are a core tool for solving many problems in neuro-science and brain science, and its development is bound to lead to innovations in the research modes and thinking patterns of important life processes.

The growth of chemical biology should serve the national strategy and satisfy national demands. China now stands at a new historical starting point. Scientific and technological innovation has been placed in an even more important position. The clarion call to develop China into a science and technology power has been sounded. Recently, China has released the *Outline of the National Strategy of Innovation-Driven Development*. It has identified the three stages for innovation-driven development: becoming an innovative country by 2020, one of the top innovative countries by 2030, and a science and technology power by 2050. People in chemical biology, together with other scientific fields, should strive to achieve the above goals and contribute to building China into an innovative country and a science and technology power!

Breakthroughs in basic research are important sources for original innovations, and core drivers for the future development of China. As an emerging discipline crossing various basic disciplines, the strategic demand for chemical biology development is, first of all, the overall improvement of innovation capability. We should emphasize the cultivation of innovative spirit, environment, and talents. Only by putting the original innovation in the first place, seizing the opportunities, overcoming difficulties and making breakthroughs in various aspects such as research tools, scientific problems, and practical applications, can China's overall level of chemical biology rank among the world's advanced countries and go hand-in-hand with or even take the lead in some frontier directions. Secondly, the convergence of multiple disciplines should be further promoted. Only through the cooperation and integration of chemistry with traditional disciplines, such as life science, medicine, and pharmacy, can we seize the high ground of science and technology, and achieve the goal that each discipline cannot reach alone in the past. In this way, we can promote basic disciplines to advance from quantitative accumulation to qualitative leap, from point breakthroughs to comprehensive improvement of system capabilities. Finally, we should vigorously advance the whole chain development ranging from basic research at the upstream to practical applications at the downstream. With a focus on the major issues related to national economy and people's livelihood, including healthcare, we should strive to translate scientific and technological achievements into clinical applications, to enable the fruits of original innovation and technological breakthroughs to serve the country and the people.

To overcome the weaknesses in the research and satisfy the national strategic demands, we should continue to give full play to the interdisciplinary advantages of chemical biology, and highlight its core position in the processes of visualization, control, and creation of living systems. Specifically, we could carry out systematic work in the following frontier directions.

4.1.1 Development of Chemical Tools for Studying the Dynamic Modifications of Biomacromolecules and Analyzing Their Relations with Human Diseases

We should strengthen the basic research on the dynamic chemical modifications of biomacromolecules (including chemical modifications of proteins, epigenetic modifications of nucleic acids, glycosylation, and lipidation) in diabetes, cardiovascular diseases, major brain diseases, cancers, and other human diseases. We can explore the rules between the modifications and the diseases, and analyze the internal relations, to provide theoretical guidance for clinical research. It is important to control the dynamic modifications of biomacromolecules with chemical and molecular means, to provide new methods and technologies for the research and treatment of clinical diseases.

4.1.2 Expansion of Novel Bioorthogonal Reaction Systems

The bioorthogonal reactions still have many flaws, such as slow reaction rates, unstable reactants, and toxic reagents. Therefore, we need to develop new biocompatible reactions that are more efficient and expand reaction types. We should aim for pioneering research achievements in the metal/nonmetal-catalyzed, light excited, or autocatalytic biocompatible reactions, and breakthroughs in reaction mechanisms, principles and applications on this basis. The funding patterns of the above areas should be forward-looking. It will be less work with better results if we actively encourage the cooperation between scientists who develop new biocompatible reactions and those tending to solve important problems in their fields with such reactions.

4.1.3 Regulation of Biosynthesis Efficiency and Development of New Biocatalytic Reactions

Great progress is being made in the scientific understanding of biosynthetic reactions, metabolic pathways, and biocatalyst mechanisms, and technologies for creating artificial biosynthesis pathways are maturing. However, efficient and highly selective applications cannot be realized in specific living systems. We should, therefore, focus on the mechanisms of limiting biosynthesis efficiency and develop relevant regulatory methods. It will greatly promote biosynthetic chemistry if strategies of chemical biology are adopted to regulate energy transfer, such as establishing an energy supply system with a specific metabolic pathway or a pathway integrated with chemical and biological means, and to drive material metabolism (biosynthesis). Additionally, we should design and optimize new biocatalysts, create efficient biocatalysts, and expand the types of biocatalytic reactions with advanced technologies, such as directed molecular evolution.

4.1.4 Development of New Technologies, Theories, and Methods

We should emphasize the combination of chemical biology and development and application of frontier technologies, such as super-resolution imaging (winning 2014 Nobel Prize in Chemistry), single molecule detection, and high throughput sequencing. Chemical biology could be utilized to increase throughputs, rates, and temporal-spatial resolutions of detection technologies, lower the limit of detection, realize single molecule detection, enhance the label specificity and manipulation accuracy of biological targets, and improve the adaptability and application scope of new technologies in special systems, including living cells, single molecules, and pluripotent stem cells.

We should further apply computational chemistry and computational biology to chemical biology, to develop theories and methods of computational chemical biology for designing, predicting, and quantitatively describing functional biomolecules and biological systems. On the basis of the enormous data generated by the chemical biology research, especially omics and high throughput technologies, we should further strengthen cooperation with bioinformatics, open up new fields of computational chemical biology, and develop theories and methods for integrating, filtering and processing these massive data, as well as for systematic analysis and prediction integrated with "drug-target-clinical effect".

The development of technologies and methods of highly sensitive detection of disease markers remains to be one of the core tasks in the field. We should promote the transition of biological marker detection from a single mode to mixed modes of various types of markers. The accuracy of disease diagnosis can be improved with the joint detection of a series of highly specific and complementary markers. In addition, we should promote the discovery of new disease markers. With the multilevel big data analysis provided by medical imaging, identification of protein post-translational modification, determination of metabolites, and other technologies, we can promote the discovery of new clinical biological markers and cancer biomarkers, and access and identify their values in clinical applications at multiple dimensions.

4.1.5 Expansion of Chemical Biology Research Directions Based on the Resource Advantages of China

Given the national conditions, and natural and human environments of China, we should promote the chemical biology research targeting our major strategic needs and resource advantages. (i) Chemical genetics research of the natural products from Chinese medicinal herbs should be highlighted. We should focus on comprehensive and in-depth studies of the target identification and the mechanisms of action of several classic natural products (such as artemisinin, triptolide, huperzine A), to make China lead in these directions. (ii) Interdisciplinary research of chemical biology and plants should be enhanced, given the important roles of agricultural production and environmental protection, as well as insufficient investment in plant research in the USA and other developed countries. We should explore the unique advantages of biomacromolecule markers, modification tools, epigenetic means and other techniques in plant research, and lose no time in carrying out forward-looking basic research, to make contributions to the key issues related to national economy and people's livelihood, such as increasing the yield of crops (rice, wheat, etc.) and preventing plant diseases and insect pests. (iii) Based on the features of an enormous population and large clinical sample sizes in China, we should further strengthen the connection between chemical biology and clinical research and diagnosis, to provide new strategies for the diagnosis and treatment of diseases, especially those having great impact on people's health (hepatitis B, lung cancer, gastric cancer, Alzheimer's disease, drug-resistant pathogens, etc.)

4.2 Conceptions and Suggestions for Further Studies

4.2.1 Further Promoting the Convergence of Disciplines

We should further unify our understanding of the interdisciplinary research at the national level and within the scientific community and fully realize its importance and urgency. As President of the People's Republic of China XI Jinping pointed out at the National Conference on Scientific and Technological Innovation 2016, during the implementation of the Strategy of Innovation-Driven Development, "we could be thrown into an unfavorable situation and miss opportunities for development— or miss an entire era—if we fail to recognize changes, respond to changes and innovatively seek change". Therefore, the communities of chemistry, biology, and medicine should all realize that the multidisciplinary research is not only a fundamental feature and a starting point of chemical biology, but also a core driver for its development and innovation. As a new secondary discipline, chemical biology plays an important role both in the growth of the disciplines of life science, medicine, and chemistry, and the development and innovation of interdisciplinary research system. The deeper its convergence and cooperation with other fields become, including biochemistry, molecular biology, structural biology, cell biology, molecular medicine, and pharmaceutical chemistry, the more obvious the research advantages are. This interdisciplinarity also promotes explorations in the frontiers of chemistry, medicine, pharmacy, materials science, and biology. Therefore, we should continue deepening the convergence of disciplines. It should be fully understood that both developments of chemical tools and research on biological problems are the source of innovation in chemical biology, and the integration of the two factors drives its sustainable development and promotes innovative research modes and thinking patterns.

4.2.2 Strengthening the Strategic Orientation of Interdisciplinarity

We should make the strategic planning and top-level design of the interdisciplinary research, chemical biology included, at the national level, and provide stable policy support. Useful attempts should be made in implementing the requirements of the Chinese government and the national outline and building China into an innovative country. In terms of the organization of scientific research, an innovation strategy of encouraging both free explorations and studies based on national demands should be applied. While promoting free explorations in the interdisciplinary fields of chemistry and life science, we should also organize large research programs on specific major scientific problems, to yield synergies in the organization of scientific tasks and to realize collaborative innovation. As to talent cultivation, we should adhere to the principle of "people foremost and academia first", actively create the atmosphere conducive for interdisciplinary talent cultivation, and enhance academic team building, to establish the development system of young talent of chemical biology covering graduates, postdocs and young researchers. Regarding scientific management and operation mechanisms, we should deepen innovation and reform to enable the synergy of science and technology and system innovations. The integration of resources and optimized allocation running through the whole chain can be realized through the pattern integrated with talent, program, and mechanism.

Research in chemical biology demands collaboration between chemists and experts from biology, medicine, physics, and other fields, to penetrate the molecular mechanism of life. Therefore, we should spare no efforts to encourage interdisciplinary cooperation, and specify measures for improving scientists' enthusiasm for cooperation. Solid support should be provided for excellent talent engaged in chemical biology so that they are encouraged to explore the unknown frontiers, carry out high-risk work in basic and original research, and work in concerted efforts for breakthroughs, which will promote original innovation and serve the national goals.

4.2.3 Deploying New Major Research Plans

The NSFC plays a forward-looking role in promoting basic research. Since the successful launch of the major research plan of "Investigations on Signal Transduction Processes Utilizing Small Chemical Probes", both the researchers and research of chemical biology in China have been significantly enhanced. It is therefore suggested to continue to deploy new major research plans concerning major scientific problems in chemical biology.

Synthesis and control of substance structure and activity at the molecular level is one of the core features distinguishing chemistry from other disciplines and an advantageous means of chemical biology for studying biological problems.

However, synthesis and manipulation of biomolecules with specific functions are still limited, compared with the target molecules with specific structures or activities, making it a research frontier in chemical biology. Correspondingly, the focus of biology will shift from the sequences and structures of biomolecules to their biological functions. For instance, "the synthesis of the human functional genome" is one of the major research topics in biological development in the near future, and another international key strategic target after the completion of the Human Genome Project at the beginning of 21st century. It is worth to note that major scientific projects of this kind can, to a large extent, promote all-round development of related disciplines, and drive a series of leaps ranging from basic research to advanced technologies. While achieving the research goals, the whole scientific community is making significant progress in various aspects, such as basic theories, research strategies, and practical techniques. Based on the comprehensive understanding of both future trends of life science and improved innovation capabilities of chemical biology, we believe that studies and interventions targeting the dynamic modifications of biomacromolecules will be an important scientific frontier and a core theme for the interdisciplinary research among chemistry, life science, and medicine.

During the life cycle, the dynamic modifications of biomacromolecules (such as proteins, nucleic acids, and polysaccharides) play a vital role in regulating cell fate. Imbalances in the dynamic regulation of key modification and/or de-modification enzymes and their substrates are important reasons for the occurrence and development of multiple diseases, including cancers. Research in this theme is both major topics and challenges in the interdisciplinary fields of chemistry and life science. They are making rapid progress and promoting the development of life science, for instance, the rise of epigenetics. However, previous and current studies are focused on the discovery of new modification types. The functions of most biomacromolecule modifications remain unclear, and there are few studies on the relations between their spatiotemporal changes (dynamic modifications) and functions. The intervention and manipulation of their dynamic modifications with exogenous small-molecule compounds are of both important value to understand physiological and pathological roles of epigenetic modifications, and great potential for disease treatment. Therefore, chemical biology will play a key role in the studies of this direction.

To promote research in this major frontier, recommends are provided. First, core scientific problems concerning "dynamic modifications and chemical interventions of biomacromolecules" as well as biological function analysis, gene expression regulation, and cell fate regulation should be identified. To study them, core research ideas and means of chemical biology, including "chemical intervention, regulation, and detection", should be developed. In future research, both the leading role of important biological problems and the application of exogenous chemical intervention should be emphasized to promote their integration. Second, in terms of tool development and application, we should highlight that chemical means and methods are universal and irreplaceable in studying biological problems. For instance, in the studies on chemical modifications and their functions of biomacromolecules, the research methods should include chemical labeling and detection (including discovery of novel modifications); small-molecule regulation, chemical simulation, and function analysis of biomacromolecule modifications; chemical modifications and synthesis of biomacromolecules, function analysis of biomacromolecules based on chemical modifications. As for research on biological problems, we should target the common problems of the life process and summarize general rules. For example, reversibility (or bidirectionality) is a shared feature of some important biological problems, including reversible chemical modifications of biomacromolecules, cell programming and reprogramming, and inhibition and activation targeting biological activity. It might be of universal research significance and thus is worthy of further exploration. Finally, we should fully realize that chemistry will be provided with a golden chance for development and innovation during the application of chemical means to solve biological problems. The chemical methods and tools developed for biological problems and demands will significantly release the research potential of chemistry to promote its update and rapid development.

Index

L. Zhang (ed.), *Investigations on Signal Transduction Processes Utilizing Small Chemical Probes*, Reports of China's Basic Research,
https://doi.org/10.1007/978-981-19-9585-9

MIX
Papier aus verantwortungsvollen Quellen
Paper from responsible sources
FSC® C105338

If you have any concerns about our products,
you can contact us on
ProductSafety@springernature.com

In case Publisher is established outside the EU,
the EU authorized representative is:
Springer Nature Customer Service Center GmbH
Europaplatz 3, 69115 Heidelberg, Germany

Printed by Libri Plureos GmbH
in Hamburg, Germany